# REMAKE: A

Imagine a better country, a better world for our children and their children. Imagine helping to build a new future, based on renewable energy, efficient transportation, and flexible, local alternatives for manufacturing and agriculture, along with improved opportunities for everyone through innovative approaches to education and higher levels of community participation.

The challenge is so enormous that the work itself is hard to imagine. Yet I believe that makers offer one of the ̲̲̲̲̲̲̲ the knowledge and skills paired with the energy and enthusiasm to lead the way. The time to start is now, to do something, however small, at home or in our local community. Together, we'll begin to make considerable progress on this giant, multi-generational DIY project, which we're calling ReMake America: Building a Sustainable Future.

In this spirit, here are some ideas to consider as first steps in the right direction.   —*Dale Dougherty*

**MAKE THINGS** » Make things that other people need. » Make things so that you don't need to buy them. » Start a business that employs people making things. » Make things closer to where they'll be used. » Repair things instead of replacing them. » Harvest usable components from devices and redeploy them. » Get to know your local salvage yard and recycling center.

**ENERGY USAGE** » Buy or build a home energy monitor that lets you see how much energy you use. » Conserve energy by understanding how much you really need to use. » Share your own energy use data. » Unplug what you don't use. » Use more energy-efficient lighting systems. » Use low-tech solutions like the clothesline when appropriate. » Take advantage of sunlight to warm your house and dry your clothes. » Experiment with generating energy at home, such as solar and wind power. » Use solar energy to heat water. » Weatherize your home, repairing or replacing old windows.

**TRANSPORTATION** » Get your bike in working order. » Invent ways to make riding bikes on roads safer for more people. » Consider how cellphones and the web can make public transportation and carpools more convenient to use. » Map the routes you travel frequently and share with others at work. » Walk when you can. » Use your car less. » Try or buy electric vehicles. » Monitor your usage of gasoline; it's a precious resource. » Drive slower to save gas. » Work from home.

**FOOD & WATER** » Grow your own food. » Experiment with different methods of growing what you need. » Cook your own food. » Raise chickens and give them your food scraps. » Know the land you live on. » Compost and create richer soil. » Use worms to create new soil. » Monitor your water usage. » Collect rainwater and reuse it in the landscape.

**LEARNING** » Learn new skills and teach others what you know. » Engage your kids, and kids in your community, in DIY projects. » Create a place to work on projects. » Embrace failure. Failure is part of learning. » Encourage curiosity and self-directed learning. » Promote hands-on projects in schools and after-school programs. » Form ad-hoc groups to share knowledge and resources.

# Make:
## technology on your time®

Volume 18

## ◢◤ REMAKE: AMERICA

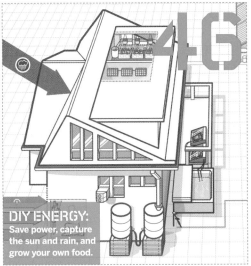

**DIY ENERGY:**
Save power, capture the sun and rain, and grow your own food.

**ON THE COVER:** The Peggy 2 LED kit will display any message you like. Here's one we like. Photograph by Garry McLeod. Styled by Sam Murphy and Ed Troxell.

## Columns

Vol. 18, May 2009. MAKE (ISSN 1556-2336) is published quarterly by O'Reilly Media, Inc. in the months of February, May, August, and November. O'Reilly Media is located at 1005 Gravenstein Hwy. North, Sebastopol, CA 95472, (707) 827-7000. SUBSCRIPTIONS: Send all subscription requests to MAKE, P.O. Box 17046, North Hollywood, CA 91615-9588 or subscribe online at makezine.com/offer or via phone at (866) 289-8847 (U.S. and Canada); all other countries call (818) 487-2037. Subscriptions are available for $34.95 for 1 year (4 quarterly issues) in the United States; in Canada: $39.95 USD; all other countries: $49.95 USD. Periodicals Postage Paid at Sebastopol, CA, and at additional mailing offices. POSTMASTER: Send address changes to MAKE, P.O. Box 17046, North Hollywood, CA 91615-9588. Canada Post Publications Mail Agreement Number 41129568. CANADA POSTMASTER: Send address changes to: O'Reilly Media, PO Box 456, Niagara Falls, ON L2E 6V2

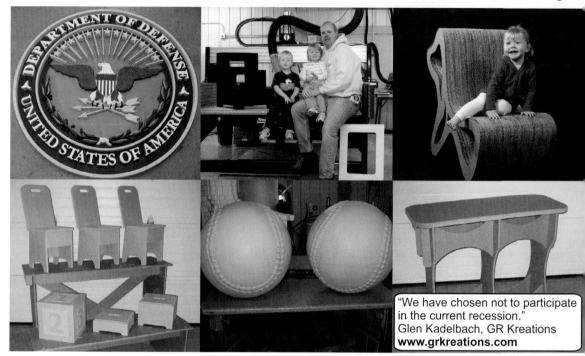

# Make: Projects

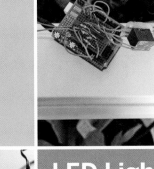
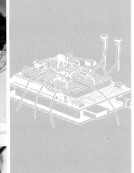

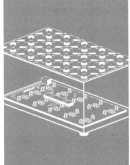

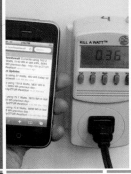

(geek)

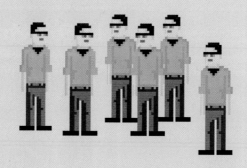

(clustergeeking)

Please geek responsibly.
You may speak the language,
but are you geeked?
Here's a chance to prove it.

# Make:
## technology on your time®

Volume 18

┌─────────────────────────────────────────┐
READ ME: Always check the URL associated
with a project before you get started. There
may be important updates or corrections.
└─────────────────────────────────────────┘

## Maker

**BIG BALSA:**
A full-sized fighter
based on the old
toy model.

# Make:
technology on your time

"We were put on this Earth to make things."
—W.H. Auden

EDITOR AND PUBLISHER
**Dale Dougherty**
dale@oreilly.com

EDITOR-IN-CHIEF
**Mark Frauenfelder**
markf@oreilly.com

MANAGING EDITOR
**Shawn Connally**
shawn@oreilly.com

ASSOCIATE MANAGING EDITOR
**Goli Mohammadi**

PROJECTS EDITOR
**Paul Spinrad**
pspinrad@makezine.com

SENIOR EDITORS
**Phillip Torrone**
pt@makezine.com
**Gareth Branwyn**
gareth@makezine.com

COPY CHIEF
**Keith Hammond**

STAFF EDITOR
**Arwen O'Reilly Griffith**

EDITORIAL ASSISTANT
**Laura Cochrane**

EDITOR AT LARGE
**David Pescovitz**

CREATIVE DIRECTOR
**Daniel Carter**
dcarter@oreilly.com

DESIGNER
**Katie Wilson**

PRODUCTION DESIGNER
**Gerry Arrington**

PHOTO EDITOR
**Sam Murphy**
smurphy@oreilly.com

ONLINE MANAGER
**Tatia Wieland-Garcia**

ASSOCIATE PUBLISHER
**Dan Woods**
dan@oreilly.com

CIRCULATION DIRECTOR
**Heather Harmon**

MARKETING & EVENTS MANAGER
**Rob Bullington**

ACCOUNT MANAGER
**Katie Dougherty**

SALES & MARKETING COORDINATOR
**Sheena Stevens**

MAKE TECHNICAL ADVISORY BOARD
**Kipp Bradford, Evil Mad Scientist Laboratories, Limor Fried, Joe Grand, Saul Griffith, William Gurstelle, Bunnie Huang, Tom Igoe, Mister Jalopy, Steve Lodefink, Erica Sadun**

PUBLISHED BY O'REILLY MEDIA, INC.
**Tim O'Reilly, CEO**
**Laura Baldwin, COO**

Visit us online at makezine.com
Comments may be sent to editor@makezine.com

For advertising inquiries, contact:
**Katie Dougherty, 707-827-7272,** katie@oreilly.com

For event inquiries, contact:
**Sherry Huss, 707-827-7074,** sherry@oreilly.com

**Customer Service** cs@readerservices.makezine.com
Manage your account online, including change of address at:
**makezine.com/account**
**866-289-8847 toll-free in U.S. and Canada**
**818-487-2037, 5 a.m.–5 p.m., PST**

**Contributing Editors:** William Gurstelle, Mister Jalopy, Charles Platt

**Contributing Artists:** Dave Bullock, Michael T. Carter, Nick Dragotta, Julian Honoré, Alison Kendall, Timmy Kucynda, Tim Lillis, Garry McLeod, Damien Scogin, Jen Siska

**Contributing Writers:** Tim Anderson, Tom Anderson, Matthew Bachler, Chris Barnes, Michri Barnes, Dan Bassak, Annie Buckley, Abe Connally, Len Cullum, Julian Darley, Cory Doctorow, Limor Fried, Saul Griffith, Andrew Haarsager, Alden Hart, Luke Iseman, Tim King, Laura Kiniry, Nance Klehm, Erik Knutzen, Tim Lillis, Terrie Miller, Forrest M. Mims III, Josie Moores, Brookelynn Morris, Eric Muhs, Windell H. Oskay, Tom Owad, Esperanza Pallana, John Edgar Park, Tom Parker, Bob Parks, Michael Perdriel, Michael H. Pryor, Celine Rich-Darley, Steven Shaw, Donald Simanek, L. Abraham Smith, Bruce Sterling, Bruce Stewart, Dave Stroud, Shaun Wilson, Adam Zeloof, Lee D. Zlotoff

**Online Contributors:** Gareth Branwyn, Chris Connors, Collin Cunningham, Kip Kedersha, John Park, Becky Stern, Jason Striegel, Marc de Vinck

**Interns:** Eric Chu (engr.), Peter Horvath (online), Steven Lemos (engr.), Kris Magri (engr.), Lindsey North (projects), Meara O'Reilly (projects), Ed Troxell (photo)

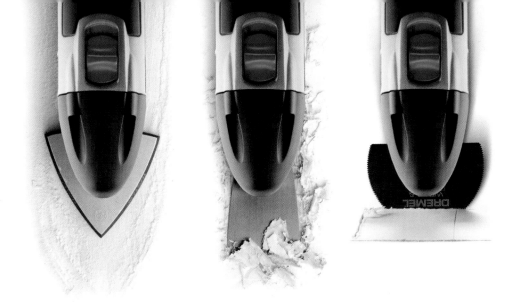

# From the company that invented versatility, here's some more.

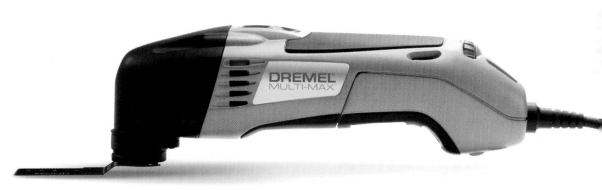

| Cut | Remove | Grind | Scrape | Sand |
|-----|--------|-------|--------|------|
| wood, drywall, metal | grout, caulk, paint | thinset, cement, mortar | adhesive, paint, vinyl | wood, metal, chipboard |

## The all-new Multi-Max™ Oscillating Tool System.

Its fast side-to-side motion and compact design make it the perfect tool for even your most demanding remodeling projects. From cutting a door jamb to removing grout—and every job in between. It's exactly the kind of versatility you've come to expect from a Dremel tool. **Call 1-800-437-3635 today for a free DVD. Or visit dremel.com to view project videos.**

**DREMEL®**
**MULTI-MAX™**
Repair. Remodel. Restore.™

Available where power tools are sold.

# Contributors

**Eric Chu** (MAKE intern) likes spinning things. For fun, he likes to throw yo-yos, build robots, and learn how to program them. He is a student at Santa Rosa Junior College in Northern California, and is taking a machining class to learn how to make his own yo-yo. Eric is the creator of Chu Pads, friction pads for bringing yo-yos back up, and is currently developing silicone response pads. His favorite food is fried rice, and his favorite color is blue.

**Alison Kendall** (Special Section illustrations) is a freelance illustrator, graphic designer, artist, and former marine biologist, currently living in San Francisco. With a B.S. in marine biology, a Master's Certificate in scientific illustration, and many years of field research experience, she decided to jump fields to pursue more creative ends. She spends her days moving vector points around and taking frequent bargain-hunting breaks to the Salvation Army around the corner. Alison's work has appeared in MAKE, CRAFT, *Scientific American*, and *Inkling*, among other publications. Check out her work at lefthandlight.com.

**Michael Perdriel** (*Off-Grid Laundry Machine*) is a Pittsburgh-based sculptor, furniture maker, product designer, and maker with an interest in off-the-grid design. He has an M.F.A. in industrial design from the University of Notre Dame. Michael is currently engulfed in renovating his home and experimenting with various green building materials in the process. He is also conducting research on aerated, lightweight concrete that exhibits good thermal characteristics, hoping to use it to build efficient wood-burning stoves for developing countries.

**Len Cullum** (*$30 Micro Forge*) is a woodworker in the Japanese style in Seattle. After working for years as a theater technician, scenic carpenter, technical director, rockabilly band road manager/effects designer/stooge, architectural dismantler, and toy designer (to name a few), it was a picture of a wooden kayak that set him on the path he continues down today. Thirteen years later, he still hasn't built that kayak. shokunin-do.com

**Abe Connally** (*On the Right Trac*) and **Josie Moores** (*1-2-3: Two-Person Shovel*) are a young, adventurous couple experimenting with sustainable technologies while raising a 1-year-old. They're currently knee-deep in the construction of an off-grid homestead, which is much better than last year, when they were knee-deep in water in the newly built first room of their home. (This was more or less a step up from living in a tent during monsoon season while Josie was pregnant.) Besides everything related to self-sustainability, they love gardening, reading, Mexican food, and concrete tools. velacreations.com

**Terrie Miller** (*Lay of the Land*) is a former online manager who left her job to play in the sun, rain, and dirt, and earned her permaculture design certificate from the Regenerative Design Institute in Bolinas, Calif. She's busy creating a miniature urban food forest in her backyard, finding ways to apply permaculture as a property renter, running permie.net, and looking for ways to increase her household pet menagerie yet still be able to travel. In her spare time, she enjoys birding, especially hawk watching, and kayaking in the Laguna de Santa Rosa. Her favorite tool is her binoculars.

# Tales of Senft Stirling engines, Einstein's Riddle, and the joys of generalism.

✉ Thank you so much for creating this great magazine. It is without question our family's Number One Favorite Periodical and it is, therefore, high time we subscribed to MAKE rather than continuously purchasing single-issue-after-single-issue as we've done since its inception!

We'll get signed up for the Digital Subscription pronto, but thought you might like to see these photos taken earlier this afternoon of our 11-year-old son, Stafford, enjoying both summer vacation *and* your latest issue! Congratulations on an inspirational and evolving project that's so well done!

—*Robin Morse, BankSide Farm, Duvall, Wash.*

✉ I saw you've started to add captions to your videos (makezine.com/podcast). That's most excellent. My wife is deaf, and there are occasions where I'd like to show her some cool web video, but a lot of times, without understanding the audio, it just doesn't make sense to her. Captions make all the difference.

Thanks for your forward thinking. Naturally, I'd expect that from the MAKE people.

—*Jared B., Portland, Ore.*

✉ Your editorials and articles on sustainability remind me of an incident that happened to a friend of mine and his wife this February.

The couple was driving home when all of a sudden their 2000 Taurus quit running. Luckily they were in sight of the local Ford dealership. They were able to get the car in to the dealer by driving in low gear. At the Ford store, the general manager said, "It would cost more to repair the car than to buy a new one."

How many times have we consumers heard that in this country over the years? This does not foster a repair culture as you suggest in your articles. This model fosters more consumerism.

Similarly, as a result of the foreclosure mess, there is a new industry created to clean foreclosed

houses to get ready to sell, and that's called "trash-out." Trash-out is where a company goes in and just guts out a house of personal belongings, with no recycling and no reuse — just fills up the dumps!

—*Allan Elgeston, Sonora, Calif.*

✉ Until I opened your steampunk issue [MAKE, Volume 17], I didn't even know that a steampunk subculture existed. With low enough costs of communication, it is easier than ever for subcultures to coalesce around their strange attractors. God bless them, each and every one.

There was a time when scientists were called "natural philosophers," and today's hyper-specialization could only be imagined as a crippling deformity on the human soul. Any thinking person was necessarily a generalist, and the entire range of human knowledge was accessible to anyone who had the desire and resources to pursue it. You could make anything imaginable because everything was made by human hands using techniques not too far removed from simple hand tools. Underlying this was the accessibility of the entire scope of art, science, and technology. Specialization? Bah! An evolutionary dead end.

—*Paul de Armond, Bellingham, Wash.*

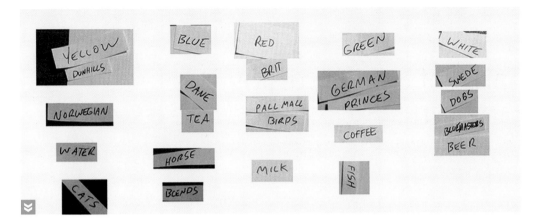

I just solved "Einstein's Riddle" in the back of MAKE, Volume 17 ("Aha: Puzzle This"). I thought I would share the creative process that ensued.

The first thing I took into account: there are 25 pieces of data involved. The second step: how do you apply logic with the data given? Knowing that to solve the puzzle I would basically be doing TRUE or FALSE variables, I devised a plan to make movable boxes with text for each of the 25 data pieces.

I wrote out each piece of data on paper and cut them out. Next I took a digital photo to bring them into Photoshop and make each piece of data its own layer. I made 5 columns, starting with house color.

The final step was to go through the questions in the same order from beginning to end. If you ask a question and answer TRUE, continue to the next question. If FALSE, adjust the tiles until the question = TRUE, then start at question 1 and repeat. If you get to the last question and it's TRUE, end the program — you just answered the riddle.

The entire process, including cutting and importing, took around 40 minutes, including 20 minutes on the actual logic. The getting there is the best part.

—James Uncapher, Plainfield, Ill.

MAKE 17 is awesome. I'm a huge steampunk fan and your team did a great job with this issue. Also, I've been catching Make: television when I can — really great job there, too.

Gareth Branwyn's article "William Blake: Patron Saint of Makers" was superb — one of my degrees is in English (the other industrial engineering), and this article hit both sides of my brain.

—James Floyd Kelly, Atlanta, Ga.

Love the magazine! What a great resource. It is important to know where ideas originate, I think, and to give credit where credit is due. People all over the world have been building low temperature differential (LTD) Stirling engines for the past 15 years, like the "Teacup Stirling Engine" from MAKE, Volume 17.

This type of engine was the result of a friendly competition between two ingenious professors, Prof. Ivo Kolin (University of Zagreb, Croatia, thermodynamics) and Dr. James R. Senft (University of Wisconsin, mathematics) starting in 1983. Kolin's first engine ran at a temperature differential less than 100°C (180°F), which was a feat at the time. By 1990, Senft's "P-19" engine, the design basis of most LTDs built today, including "Teacup," could run on a temperature difference, hot side to cold side, of 0.5°C — less than 1°F!

For this reason, it has been suggested (baileycraft.com/senft.htm) that these engines be referred to as Senft Stirling engines. And yes, there have been consumer applications proposed (makezine.com/go/senft). These engines are described in loving detail in Dr. Senft's book, *An Introduction to Low Differential Stirling Engines* (Moriya Press, 1996).

Thanks again for a wonderful magazine. Even better than the *Popular Mechanics* of my youth!

—Bill Dreschel, State College, Pa.

## MAKE AMENDS

*A handful of errors made it into MAKE, Volume 17:*

In "The 'Discreet Companion' Ladies' Raygun" we regrettably misspelled the name of photographer John Keatley.

In "The Florence Siphon Arabica Brewing & Extraction Apparatus," on pages 65–66 all references to 4mm tubes and holes should be 7mm.

The name of designer Silvia Bukovac Gašević was misspelled on page 90. We regret the error.

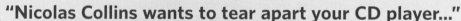

# Announcing: Makers Market

Ever since MAKE's first foray into e-commerce three years ago, we've believed an important part of our mission should be to help indie makers bring new DIY kits and products to market.

Breaking into a market through traditional retail channels can be a daunting barrier for an indie maker or small kit builder.

That's why in the Maker Shed (makershed.com) we're willing to carry small runs of promising new project kits, often assembled by hand in the maker's garage and packaged in small, nondescript boxes purchased at the local packaging store. We do business over email with a virtual handshake; put the product online; maybe blog it on makezine.com to stir up interest; watch the sales and listen to what customers have to say; and then give the kit builder valuable feedback that he or she can use to refine the idea. It's another way MAKE reinvests in the maker community.

But makers' product ideas and the resulting economics are as varied as the makers themselves. They just don't all fit in the Maker Shed. Not a week goes by when I don't have to take a pass on some cool kit or game concept.

Sometimes, they're actually services or custom one-off products that would be difficult to carry in the Maker Shed. Sometimes they just don't have enough margin to split between the maker and the retailer.

Nonetheless, all these makers and their DIY creations need a market to showcase their ideas and their products.

And so, this summer the Maker Media team is joining with the Boing Boing crew to launch a brand-new online marketplace called **Makers Market** (makersmarket.com).

Makers Market will be a curated marketplace where indie makers, small suppliers, and artists will be invited to sell their creations, and in some cases their services, to the DIY community.

Participating makers will have their own store-front or studio space where they can showcase their work and sell their products, have their own blog,

> We're setting up a curated bazaar of wonderful science, tech, and artistic treasures. Join in!

post pictures and videos, and communicate with their customers and the DIY community at large. Plus, we have a few other cool marketing tools we'll be integrating.

Think of it as a curated bazaar of wonderful science, tech, and artistic treasures. We provide the web service, the tools, and the community. Participating makers will be responsible for doing their own product fulfillment and for bringing their unique character, energy, and DIY spirit to the marketplace. And in exchange, the Makers Market takes a small commission on resulting sales.

We'll be providing more Makers Market updates and details through makezine.com, craftzine.com, and boingboing.net as we get closer to launch later this summer.

If you're a maker of original products (anything from Arduino shields to blacksmithed jewelry to catapults) or have a service of interest to other makers, we'd love to hear from you. Bonus points if you're willing to actively engage the community from within your Makers Market storefront.

## » INTERESTED?

Send us a note (makersmarket@makezine.com) telling us who you are, what your product or service is, and why you find the Makers Market intriguing. If you have an existing store, that's perfectly OK. Send us a link. If you don't, that's OK, too. Now you will.

Dan Woods is associate publisher of MAKE and general manager of the Maker Shed.

# If You Can't Open Government, You Don't Own It

On his first day in office, President Obama embraced the MAKE ethic in the most meaningful way: by opening the government. Since the infamous Ashcroft memo, issued after September 11, 2001, the U.S. government had treated the Freedom of Information Act as a bug, not a feature. FOIA is the legislation that lets ordinary citizens and corporations ask government agencies to disclose the contents of their files.

You might want to know how the EPA decided that the old Superfund site your kid now plays on was safe for human habitation. You might want to know whether the contractor who supplied your body armor had taken out any big-shot generals for a steak dinner … on a private jet, en route to a Caribbean island.

This is what the FOIA is for, and when it works, it makes government accountable and more secure, rooting out corruption and inefficiency through the time-honored method of sunshine and lots of it.

John Ashcroft, then U.S. attorney general, issued a directive to government agencies on Oct. 12, 2001, that gutted FOIA. Under the new directive, agencies were advised to deny FOIA requests, unless there was a "sound legal basis" for complying with them.

Prior to this, agencies had defaulted to honoring all FOIA requests, unless there was some "foreseeable harm" that could come from them. Effectively, Ashcroft changed FOIA policy to: "We'll honor your FOIA request — after you win a lawsuit against us."

So it was a grand and exciting day for activists of all descriptions when, on Jan. 21, 2009, President Barack Obama issued a memo reversing this policy, directing government agencies to "adopt a presumption in favor of disclosure" — that is, to change the government's default position on revealing what it's doing from "None of your business" to "Pull up a seat and let me tell you all about it."

MAKE was founded on the principle that "If you can't open it, you don't own it," the stirring preamble to Mr. Jalopy's infamous Maker's Bill of Rights. This is even more true of governments than it is of gadgets. Governments do business on our behalf, with our money, in our country. There's never a good reason for the government to keep its everyday workings a secret from the people who own it: the citizenry. Secrecy breeds waste, corruption, and insecurity.

President Obama went even further: the Jan. 21 memo tells agencies that: "They should not wait for specific requests from the public. All agencies should use modern technology to inform citizens about what is known and done by their Government. Disclosure should be timely."

What's this mean for you? Well, it means that government is in the open data business. From now on, the daily workings are supposed to be an open book. That's fine news indeed for makers: time to get cracking on the services and systems that make that mountain of data into something meaningful.

Americans could do worse than to look at their British cousins, who, through online services like TheyWorkForYou (theyworkforyou.com), have made a delightful nuisance of themselves by slicing and dicing their government's records. Just feed in your postal code and out pops your rep's name, every word she's uttered, contact details for neighbors of yours looking for support on lobbying her, a one-click way to send her an email, tallies of how often she votes against the party line, and an RSS feed for every word she speaks from here on in.

It's a measure of just how scared governments get of this kind of thing that the British Parliament tried to pass a law saying that its members didn't have to publicly disclose the details of their expense reports — that they could keep their spending of public money on personal expenses a secret. Of course, services like TheyWorkForYou helped clobber this initiative by making it trivial for Britons from across the country to send "Are you kidding me?" emails to their Members of Parliament, shocking them into action. Needless to say, the bill failed.

President Obama's inaugural address lionized "the risk-takers, the doers, the makers of things." Last time I checked, that was us.

---

Cory Doctorow lives in London, writes science fiction novels, co-edits Boing Boing, and fights for digital freedom.

Photograph by David King

Photograph by Michah Gibson

# From Garbage to Gallery

"You can basically find anything you want at the dump," says artist **David King**. He would know. He spent 20 hours a week sifting through junk at the San Francisco city dump as part of a four-month artist residency. "A lot of the stuff is still usable, and some of it is even new and still in its wrapper," he adds.

An important part of King's residency was to share the recycling experience with children from local public schools. The classes that came through his studio definitely shared his enthusiasm for making art out of garbage, making their visits "the highlight of the residency," he says. "It's great to get kids excited about using trash to make fun stuff."

The fun stuff that King made includes numerous spherical sculptures that mime scientific forms — cells and viruses, satellites and planetary bodies — but that also evoke the playful sensibility of their raw material: sports balls.

Upon discovering numerous discarded basketballs, baseballs, tennis balls, golf balls, and more, King decided to use them as structural elements. Then he attached smaller found items to the balls, such as hair curlers, screws, game pieces, and suction cups.

King had previously explored circular forms through collage, and these ideas only grew with his new finds. "I came to the residency with some pre-conceived ideas about what I would do there, and what I would find in the waste stream," he says, "but I quickly had to give all that up and just respond to what I found there."

For their part, the schoolchildren had plenty of ideas. "Many kids would get so excited they could hardly finish their sentences when they wanted to tell me their suggestion for some of the materials I had collected," King explains. "Two fifth-graders that came through, already looking like art school students, told me that making art out of trash was their dream job. When I told them I also received a $1,000-a-month stipend, they went through the roof."

Uniting artists and kids to bring a creative edge to recycling? As the fifth-graders said, "Awesome! Cool!"
—*Annie Buckley*

≫ **King of the Dump:** davidkingcollage.com/dump1.html

# On the Right Trac

Farm equipment can be expensive and hard to maintain, but open source enthusiasts are out to change all that. In rural Missouri, **Marcin Jakubowski** and the team at Open Source Ecology (OSE) are designing a sustainable village for the future. One of their latest prototypes is an open-source tractor, called LifeTrac.

Inspiration for the project came from disappointing experiences with used tractors. After spending $2,000 to fix a transmission on an old Massey Ferguson that died a week later, the team received a donated Allis-Chalmers D-17. After a bit of light work, the clutch on the Allis went out. Instead of spending $1,000 to fix that, Jakubowski decided to explore another option: make their own, better tractor. "I liked $100/year maintenance costs more than $2,000/year costs," Jakubowski explains.

Spending time when and where they could, the team put the prototype together in four months at a cost of $4,000. LifeTrac comes complete with a hydraulic drive system, front-end loader, articulated steering, a 55hp diesel engine, and 4-wheel drive to boot. They've also added backhoe, PTO generator,

and rototiller attachments to make the tractor even more practical.

The team has used LifeTrac for various projects around their Factor e Farm, located an hour north of Kansas City, Mo., and Jakubowski says the results have been outstanding. "We used LifeTrac to pull the Allis out; friends hauled it off. That's a political statement on OS equipment replacing design-for-failure-and-high-cost equipment."

Next for the team: more attachments for LifeTrac, and a small, walk-behind version called MicroTrac, with its engine and other parts interchangeable with the big version. The team is also working on a solar steam engine, a hydraulic compressed-earth block machine, a sawmill, and modular construction techniques and materials. OSE is currently accepting donations from dedicated fans to help continue their efforts with the sustainable village model.

—*Abe Connally*

≫ **Open Source Ecology:** openfarmtech.org

**LifeTrac Build Videos:** makezine.com/go/lifetrac

Photograph by Marcin Jakubowski

Photograph by Christian Nicolson

# Big Balsa

When New Zealand artist **Christian Nicolson** was a kid, he played backyard World War II games with a model of a British Royal Air Force Spitfire, the sort of cheap, fun toy you'd press out of a single sheet of balsa wood. They often ended up stuck in a tree or shattered on impact.

Now fully grown, Nicolson wanted to represent his childhood fantasy of piloting a Spitfire through his art. The result is a full-sized fighter based on the old balsa models. Like the hand-launched toy, the sculpture is printed on one side only. The nose shroud may be made from lead rather than Plasticine, but the look is true to form.

Nicolson built the Spitfire in late 2008, throwing every spare dollar into the project. He also recruited volunteers, paid only in beer, to help out. "I called in all the favors I could. The problem was that even to turn the wings over, I needed a hand," he says.

Clad in 6mm (¼") Fijian kauri marine plywood, the sculpture is built with steel bracing and a polystyrene interior to keep the weight down. Nicolson says pre-production was vital, and a lot of time went into research. "It had to be rigid; I didn't want the wings to droop," he says.

Coming in at 1.3 metric tons (1.4 tons), it may be too big to throw, but it's also too solid to break. It's 10.4 meters (34 feet) long with a wingspan of 11 meters (36 feet). It comes in five sections that bolt together, and when disassembled it fits easily onto a domestic trailer.

Prior to building the Spitfire, Nicolson built a wingtip and rear fuselage for a "crashed" Japanese Zero fighter sculpture, made from laminated sheets of macrocarpa (aka Monterey cypress). Tying the two ideas together, the Spitfire bears a small "kill" flag suggesting that it was the plane that shot the Zero down. "Some make-believe play going on there," he explains.

The Spitfire has been shown at a couple of New Zealand art exhibitions, and Nicolson is now seeking a buyer. Ships flat!

—Steven Shaw

>> **Spitfire Replica:** christiannicolson.co.nz/spitfire.html

# Trailer (Re)Made

Inspired by post-Katrina New Orleans, New Yorker **Paul Villinski** created Emergency Response Studio (ERS), a FEMA-style trailer transformed into a fully functional, sustainably built mobile artist studio. His vision: a healthy space for visiting artists to embed themselves in post-disaster settings, and a prototype for what temporary housing can become.

Villinski purchased the 30-foot Gulf Stream Cavalier through the U.S. Goverment Accountability Office (GAO), the same agency that sold toxic FEMA trailers as provisional shelter after Katrina. Then, for seven months, he worked at gutting and re-envisioning the trailer using eco-friendly materials donated by more than a dozen sponsors.

The 48-year-old artist says that integrating mobility into his artwork is a natural progression. "I grew up an Air Force brat," Villinski says. "We were always moving, road-tripping, packing things up … the gypsy life is in me."

The result is a bright, inviting space that incorporates rather than excludes its surrounds. Inside, he insulated the ceiling with recycled pre-consumer denim scraps; crafted the kitchenette from Plyboo, a bamboo sheet material; and insulated the walls with mineral wool derived from blast furnace slag. While the trailer's reclaimed plywood wall paneling isn't green in the traditional sense, Villinski says it's probably more sustainable because it was recycled locally.

Some of ERS' most innovative features include a crank-down wall that lowers into a sturdy deck and expands floor space; a sub-floor encasement holding 1,300 pounds of batteries that store energy made by rooftop solar panels; and a geodesic skylight expanding headroom from 6½ to 10 feet. There's also a tilt-down mast that supports a wind turbine, the original version of which Villinski built using MAKE's instructions (see Volume 05, page 90).

After touring Texas this spring, ERS is headed for Wesleyan University in the fall, but not before catching some rest. "Believe it or not, I've never spent the night in it," Villinski says. "I'd like to this summer."

—*Laura Kiniry*

≫ **Response Studio:** emergencyresponsestudio.org

Photograph courtesy Jonathan Ferrara Gallery, New Orleans

# Wood-Grained Visions

Three adolescents cruise the desert, two holding skateboards while another steers a bike. It's a familiar scene rendered in an unexpected way by Brooklyn-based artist **Alison Elizabeth Taylor**. Rather than loosely sketched watercolor or street-inspired airbrush, Taylor uses marquetry, or wood inlay, a meticulous process typically used for decorative purposes and popular during the Renaissance.

A time-consuming technique, Taylor's marquetry involves cutting and shaping small bits of wood, arranging them precisely, and securing them with resin and a vacuum press. Taylor uses wood veneer as others might use paint, expertly choosing each piece so that the color and grain lend depth and form.

For a 2008 exhibition at James Cohan Gallery in New York City, she constructed a room-sized environment entirely from carefully fitted pieces cut from more than 200 different types of wood.

This ambitious project had a daunting wow factor, yet the individual pictures resonate like dreams. In *29 Palms*, the teens and the surrounding landscape are depicted in a seemingly endless array of browns, their forms defined by alternately striated and wave-like patterns of wood grain.

Taylor's knack for composing is cinematic, as if black-and-white movies were updated and remade with puzzle pieces. The bikini-clad ladies in *Swimming Pool* seem caught in the middle of action; one hand draped around a martini glass could have leapt from the screen of any number of movies. In *Era of Argus*, a tattooed man kneels to feed a peacock, its tail feathered with glistening browns. A pair of hands reaches eerily from a sea of dark wood, grain swirling like water, in *Hands*.

Pairing an age-old practice or material with newfangled ideas has become de rigueur for many young artists, and Taylor does it with elegant flair and precise workmanship. Through the monochromatic lens of a wooden palette, she expresses loose narratives that often take on mythological dimensions.

—*Annie Buckley*

≫ **More Wood Work:** jamescohan.com/artists/ alison-elizabeth-taylor

# ¡Luchadora Libre!

Like many artists, Los Angeles-based **Sophia Allison** started out painting portraits. But a fascination with WWE wrestling eventually led this painter down a far more unconventional path.

Starting with a wrestling mask she already owned, she says, "I turned the mask inside out and studied how it was put together. I created a rough pattern, and worked off of that to create the pieces for my first homemade mask."

Allison's interest in wrestling led her to *lucha libre*, Mexico's version of the sport, which relies heavily on the power of masks to alter and transform the identities of *los luchadores*. Still, something was missing.

"I was disheartened that there weren't many well-recognized female wrestlers — those that were known in American wrestling were seen primarily as sex symbols," she says. "This sparked my interest in creating masks and capes that fit me, and these sort of became alter egos."

So how does a painter learn to design and stitch sculptures that double as wearable art? (Witness

her chic, pink wrestling ensemble with intricate embroidery and interior padding stitched from individual maxi pads.) In 2002, Allison bought a sewing machine and taught herself to sew.

Since then she's created a whole retinue of alter egos, made from domestic or recycled materials found around the house. Band-Aids, shoelaces, and various felts and fabrics have found their way into her work. One cape is hand-sewn from used tea bags, while some tough-looking headgear is crafted from deconstructed Converse basketball shoes.

A vermilion felt mask, dripping with shimmering pink threads, packs a punch. A flesh-toned, tapestry-and-nylon mask has an eerily menacing appearance, part ghost and part hosiery-horror-show.

Although Allison's imaginatively feminine wrestling attire hasn't yet entered the ring, it provides endless possibilities to transform the contemporary Everywoman into an explosive domestic diva — Supergirl, beware.                     —*Annie Buckley*

≫ **Luchadora Masks:** sophiaallison.com/maskindex.html

Photography by Chloe H. Park

# Solar Asphalt Gondolier

Inspired by the bamboo rickshaw on *Gilligan's Island*, cyclist **Reno Tondelli, Jr.** not only built "the world's first recumbent hybrid taxi," he took the leap and turned it into a pedicab business that now supports his family.

Tondelli, 43, powered up an Organic Engines recumbent trike with a 600-watt DC motor and six 12-volt AGM wheelchair batteries. To handle the extra weight, he welded new dropouts, built new wheels with fat 10-gauge spokes, and added beefy tandem hubs. He tricked it out with a stereo, aluminum fenders, and vinyl canopy, then topped it off with trickle-charging solar panels and a squeeze-bulb horn.

Two years and $8,000 later, he had the world's first four-passenger, zero-emissions, solar-assisted pedicab. He dubbed it the Solar Gondola and made extra cash and lots of smiles pedaling tourists in his hometown of Venice, Calif., while sporting a handlebar mustache and straw boater hat. "The ride is phenomenal," Tondelli says. "It's smooth, it's quiet, it's fast, comfortable — and a wee bit stylish!"

Style points don't count with the cops, however,
and the asphalt gondolier was frequently hassled for not being street-legal. In 2008, Tondelli relocated with his wife and son to Colorado, where the solar bike is now a legitimate taxi with seatbelts, insurance, and a tractor spotlight for nighttime use. He works out of a shop shared by pedicabbers and horse-drawn carriages in downtown Denver.

The film industry refugee worked 20 years in the movie biz "making stuff look cool." Besides taking a welding class, he mostly learned just by doing it.

Upgrades are in the works. He's eyeing new solar films, a BionX motor that recaptures braking energy, and long-life lithium batteries. But is this a world-changing technology? Nah, says Tondelli. "The recumbent pedicab's a lot better on your shoulders and neck than the upright. Business-wise? It's a wash. But I am the one-and-only solar pedicab — it brings a lot more frivolity and gaiety to the world."
— *Keith Hammond*

>> **Tondelli's Solar Gondola:** solargondola.com
➕ **More Photos:** makezine.com/18/gondola

Photograph by Bill Manisculco

# 5,000 Days? $CO_2$ Targets and How Much Fossil Fuel We Can Burn

A recent Gallup poll shows that 41% of Americans think that reports of climate change are exaggerated, a number that's grown in the past few years. For someone who has spent a lot of time thinking about climate change and the best way to approach it, this is pretty depressing news.

In order to combat misinformation, it would be really useful to know the urgency of action on climate change: is it needed today, tomorrow, or in 2020?

We often hear the necessary action for climate change expressed as a percentage reduction by a certain year. The current consensus goes something like this: "We need 80% reductions by 2050." This obscures a whole lot of very important details. Let's start with the simple one. What's implied by these statements is that we need to reduce our energy use by 80% of 1990 levels of global $CO_2$ output. That sounds straightforward, but is that the reality of climate science?

Not really. The problem with expressing the $CO_2$ reduction target as a percentage is that it hides important facts — that we know reasonably well how $CO_2$ ends up in the atmosphere, and that $CO_2$ has a long residence time in the atmosphere — and suggests we'll somehow be able to continue to emit some amount of $CO_2$ in the future and still be OK.

In the groundbreaking book *Energy Policy in the Greenhouse*, the authors expressed the $CO_2$ problem in a more honest way. They calculated how much the $CO_2$ concentration in parts per million (ppm) increases for a given amount of energy consumed from different fossil fuel sources. Their numbers are still quite accurate in describing humanity's influence on $CO_2$ concentration in the atmosphere:

**1 billion tons of carbon = +0.260ppm $CO_2$**
**1TWyr of coal = +0.198ppm $CO_2$**
**1TWyr of oil = +0.155ppm $CO_2$**
**1TWyr of gas = +0.112ppm $CO_2$**

A TWyr is a terawatt-year, or 1,000,000,000,000 (1 trillion) watts for 1 year, or $3.1556926 \times 10^{19}$ joules. Given that the world uses more than 10TW of fossil fuels, which means 10TWyr of energy each year, you can see how it adds up quickly. Remember that $CO_2$ is at about 387ppm today. The above relationship shows that the way we do things today, we add roughly 1ppm–2ppm every year.

We can turn the analysis around here in a manner that's really interesting. Averaging the $CO_2$ output for each of the different fossil fuels, we can now state that for each joule of energy we get from them, we produce a $5 \times 10^{-21}$ ppm increase. This is a tiny, tiny number, but we use a lot of joules. Producing a single can of Coke uses around $5 \times 10^6$ of them. In this way, we can estimate the effect of billions of small actions.

But to return to the question at hand: how soon must we act? We need two more pieces of information. What temperature do we want to stabilize at? And what $CO_2$ ppm does that correlate to? Warming of 2°C (3.6°F) above preindustrial levels is considered something of a "point of no return" by climate scientists, beyond which we will see very negative consequences. Two degrees implies 450ppm as an upper limit. But that might still be too high, given the even more ambitious 350ppm prescribed by Jim Hansen of NASA and other leading climate thinkers.

So let's say we accept the risks of 450ppm (and I mean risks — this only gives us a small chance of staying below 2°C of average surface warming). What we get from the above equation is that we only have about 400TWyr left, or 40 years burning at 10TW. Or only 20 years burning at 20TW.

And that's only the first bit. We also need to decide how much power we'd like from new sources of non-carbon energy that don't exist yet. If you said you'd like the world to use the same amount of power as it does today in the future (this ignores population growth and growing demand, hoping that efficiency measures offset that), then we'd need 16TW of power, around 12TW of which come from fossil fuels.

Why do we need to know this? It's to figure out how many ppm of $CO_2$ will be added to the atmosphere to create our new energy-generating and energy-using infrastructure. What I mean specifically is that at least for a while, we'll be using coal, oil, and natural

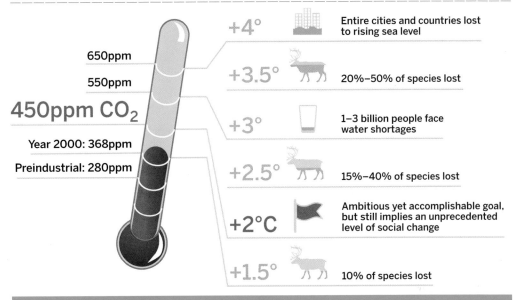

| | | |
|---|---|---|
| +4° | | Entire cities and countries lost to rising sea level |
| 650ppm | | |
| 550ppm | +3.5° | 20%–50% of species lost |
| 450ppm CO$_2$ | +3° | 1–3 billion people face water shortages |
| Year 2000: 368ppm | | |
| Preindustrial: 280ppm | +2.5° | 15%–40% of species lost |
| | +2°C | Ambitious yet accomplishable goal, but still implies an unprecedented level of social change |
| | +1.5° | 10% of species lost |

**CAUSE AND EFFECT:** Even if we hit our target of atmospheric CO$_2$, there's uncertainty about what temperature the Earth will eventually equilibrate at. This uncertainty generally implies that we should be conservative. I choose 450ppm because it's a very ambitious goal (much more ambitious than the 550ppm chosen by the Stern Report, for example) yet it still implies a high level of doom and gloom, and an unprecedented level of social change.

Illustration by Kirk von Rohr; sources: "Avoiding Dangerous Climate Change" (www.stabilisation2005.com/outcomes.html) and IPCC 4th Assessment Report (www.ipcc.ch/ipccreports/ar4-wg2.htm)

gas to create the new solar cells, electric cars, wind turbines, nuclear power plants, green buildings, and mass transit solutions, until we can make those machines with clean power.

Using conservative assumptions for how much energy it will take to create these things, you see that it will take:

» 0.5ppm of CO$_2$ to completely replace the world's fleet of around 1 billion cars with small, light, 1,000kg electric vehicles
» 6ppm of CO$_2$ to make 5TW of new solar generating capacity
» 0.5ppm of CO$_2$ to make 3TW of new wind power
» 0.5ppm of CO$_2$ to make 2TW of new geothermal power
» 9ppm of CO$_2$ to make 250 million new "green homes"

So that's about 16ppm already. Add some nuclear plants and some biofuel refineries, some wave power machines, some new rail, and quickly you'll see we can easily expend 25ppm of CO$_2$ just building the new infrastructure we need for a new economy. That would take us to about 415ppm from where we are now. Then we've only got 250TW years of fossil fuel burning left. That's 20 years at the rate we're burning today.

If we can commit to a CO$_2$ ppm target, we can figure out how to get there and how many fossil fuels we've got left to use. If we stick with the percentage model, we're flying blind. This kind of "working backward" analysis is what businesses do when they set a strategic goal, and it's what we need more of if we're to solve the climate problem once and for all.

So if you don't want to fly blind, and you want a strong target so we actually get to where we want to go, you have to answer a few of the questions above. If you say we've got 5,000 days left until we have to act decisively, you're playing a very risky game. After all, 5,000 days is 13.6 years. If we do nothing for 13.6 years, and then we make the decisions above, we only get 7 years of our current energy consumption before we don't have any more energy to run humanity's needs. Every joule after that has to be put toward the new infrastructure.

I know there are a lot of crude assumptions here, so feel free to tear those apart; the two main points are that we need to commit to hard numbers of where we wish to go, and we need to consider carefully the cost of building infrastructure, because we only get one shot at this.

Saul Griffith is a co-author of *Howtoons* and a MacArthur fellow. saulgriffith.com

# Designer Futurescape

The tectonic moment — for me, anyway — came when Anab Jain, a young interaction designer educated in India, Vienna, and London, stood up in front of seven hundred Web 2.0 zealots in Geneva and declared that she works in "design futurescaping."

I know that sounds plenty weird, but since this is MAKE, I can assume that I'm addressing contemporary grown-ups here. If you've got the jazzy MAKE Digital Edition rather than the inky real-world paper mag, just click the ol' pull-down menu, Google "Anab Jain," and in five minutes you can discover that this inventive personage, who is about as global as a dandelion seed, really does work in "designer futurescapes." I would challenge you to find any better description for what she's up to. Really: I'll wait.

Still with me? OK, now imagine you are at this same event in Geneva — the Lift conference, they call it — and a guy named Matt Webb grabs the stage and describes his design work, too. He says he uses design techniques to "walk the landscape of possibilities." Webb deploys market research, economic analysis, prototyping, physical models, user observation, and historical parallels, but mostly — so he says — he uses stories: "The story is your laboratory, reading is your research, and writing is your experiment."

I can't tell you how disconcerting it was to be a career science fiction writer and hear that publicly declared by Webb. Mind you, he is clearly not a lunatic. The guy's a robotics hacker and a cognitive psychologist, and he writes for O'Reilly Media publications, clear signs that he's no crazier than anybody else in this magazine.

If you Google "Matt Webb" — of Schulze & Webb — you'll discover that when he claims he's a designer walking landscapes of possible worlds through his storytelling, he's (a) not kidding and (b) not doing science fiction. This is what the guy does when he gets out of bed in the morning, as an *industrial practice*.

How did we stumble into the glimmering dawn of a "designer futurescape"? Maybe that's got something to do with "virtuality." People who do web design are used to products and services that are vaporous and speculative, stuff like Vinton Cerf's

## Handheld internet machines are reshaping cities in the way cars reshaped cities in the 20th century.

Interplanetary Internet, which, by the way, ol' Vint just booted up, so it's real now. But I don't believe that's what has happened — because this is 2009, and virtuality is a corny early-90s paradigm.

We're entering a new and different situation. Handheld internet machines are reshaping cities in the way that cars reshaped cities in the 20th century. When our cities — our real places, you know, glass, bricks, roads — become the products of "urban informatics," why would any normal guy call that reality "virtual"?

Get in a car with a GPS unit and look at what that does to what we innocently called "time and space." Maybe that sounds like a Google-map detour from my point here, but imagine Webb 20 years from now. Graying at the temples, our Matt explains to normal people that he "walks landscapes of possibility" for a living. Well, his listeners are "walking landscapes of possibility" to get down to the grocery to pick up a head of cabbage. Come 2025, 2035, and there's nothing rhetorical or far-out about this. It's become "real life"; it's natural everyday behavior, like punching a button on a wall and seeing light pour out of the ceiling.

Now, I won't claim that this is technologically inevitable, because that kind of linear technological determinism is a dead form of futurity. As Webb said — rather insightfully, I thought — society, human nature, and technical artifacts are a tightly linked trio, like pressure, temperature, and volume.

Turn up the heat with some techie gizmo, and society and individuals will kick back tout de suite. So we've got a networked, interactive, increasingly speculative futurity. It's got user-centric Google maps rather than officially certified paper road maps. It's not some Marxist road to utopia, it's a navigable global sprawl.

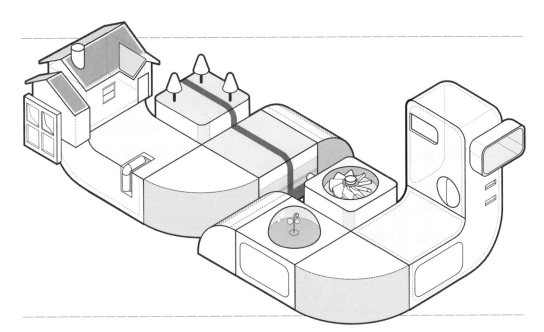

Illustration by Julian Honoré/p4rse.com

You might argue — as a hardheaded skeptic — that design in America and Europe is being shoved into these frothy conceptual spaces because the Chinese dominate the real-world heavy industry. That's the lesson I took from Lift speaker Jörg Jelden, a "trend analyst." Yes, that's his real-world day job.

The Chinese have become the workshop of the planet. They make heavy-industry factory products that are physical and "real" — you can drop 'em and break your foot. Very real, right?

But a hell of a lot of those Chinese products, something like a quarter or a third, are "not real" — because they're fakes. Jelden claims, and I'm afraid I believe him, that some of these "fake products" are not even cheap mimics of better ones. No, today's offshored fake products can be more innovative than real ones because pirates evade the limits of patents, copyrights, and government regulation. Try shoving back that tide, RIAA and MPAA.

Worse yet, those fakes often emerge from the same factories that make the branded, certified real stuff. The really clever Chinese operators are salting their shipping containers with a profitable mix of "real" and "fake," rather like adulterating the baby food with melamine.

Clearly this is sinister activity, and I don't want to valorize it, but I hope it shows that this phenomenon is not just something that wacky designers are blue-skying about in Geneva. We've got a black-global criminal underclass of hard-bitten characters who methodically chew away at the "real." And they've got more designer futurescape to wander around in because they don't have to fret about intellectual property.

I could go on about this issue all day — I may have to go on about it for 10 or 20 years, because it's really bugging me — but I want to end by imploring you to think about what this means for makers.

Is it really so important to make a real thing, to have an empirical referent, a "real" physical object that actually sits there, solid on the mantelpiece? Or is the true action in distributing the potential to make things through a webbed community of makers?

Not "virtual reality" — that's so old-hat — but "reality as a web service." A real landscape full of real stuff, that is mapped, infiltrated, subsumed, by a futurescape full of potential stuff.

Distributed, collaborative reality in permanent beta. That describes the futurescape, and it makes the former distinctions between designers, science fiction writers, trend analysts, and even manufacturers look quite archaic.

Maybe that sounds abstract, far-fetched and freaky, never so rock-solid and reassuring as physical real estate. But just try to price some "real estate" now. Where's the beef there, where's the reality? The Earth's a big place — so find some GPS spot where today's battered landscape escapes the shadow of the futurescape.

I don't think we have any such place left now. What's more, I don't think we have any way to get back there. It's futurescape right, left, and center, north, south, zoomable, and all around. We made the futurescape, and now we've gotta use it.

---

Bruce Sterling (bruces@well.com) is the author of several science fiction novels and nonfiction books.

# Living the Good Life

What we can learn from a Depression-era homesteading couple. By Matthew Bachler

In 1932, at the height of the Great Depression, Helen and Scott Nearing packed up their life in New York City and settled on a 65-acre farm in the Green Mountains hamlet of Winhall, Vt., where they set out to establish a "self-sufficient household economy." Over the course of the subsequent 20 years, the Nearings converted a deteriorating farm into a materially productive and spiritually rewarding homestead, a story encapsulated in their now classic 1954 work, *Living the Good Life*.

It seems historically appropriate to be discussing the Nearings' story once again, on the precipice of an economic downturn that may very well rival the severity of the Great Depression, one impetus of the Nearings' urban emigration. Along with the protracted depression, the Nearings' outspoken socialist politics resulted in their rejection of — and

also by — a staunchly capitalist status quo. Over the previous three decades, government and business forces had responded with intimidation and violence against a burgeoning labor and Socialist movement. The Nearings found themselves unable to teach or publish. This was the era of Emma Goldman, Eugene Debs, and the Wobblies; at no other time in American history had capitalism been so pushed back on its heels.

In establishing their homestead, the Nearings had three main goals: to remain solvent and as independent as possible from the larger economy; to cultivate and adhere to those values they saw as being healthy and essential to the good life, namely simplicity, freedom from anxiety, harmony, and purposefulness; and to schedule leisure time into every day so as to encourage social and personal improvement.

Photography from the Scott and Helen Nearing Papers at the Thoreau Institute at Walden Woods

Further, they laid out a 12-point, ten-year plan that functioned, in their words, as the "Constitution of our household organization." Its most notable principles were a rejection of an exchange-value economy premised on profiteering and an embrace of a use-value one, and a commitment to carry out operations on a "cash and carry basis" independent from a capitalist banking system.

After outlining the principles of the good life, the Nearings walk readers through their day-to-day practices of homesteading that allowed those principles to be met. On home construction, they advocate building from stone, as stone homes are more harmonious with their natural surroundings, materials can easily be sourced locally, and they are naturally cooler in summer and warmer in winter.

In food production, three obstacles faced the Nearings. The mountain valley in which they settled had only 85 reliably frost-free days a year. In response, vegetables and fruits were divided between two gardens: one located close to the home, where frost-tolerant plants were placed, and a second "insurance garden" planted on higher ground, where temperatures were slightly warmer. The pitch of their land was quite steep, so the Nearings built a complex system of terraces. Finally, the soil was depleted of nutrients, which they reversed over years of tilling in homemade compost.

When it came time to eat the literal fruits of their labor, the Nearings followed a strict vegan mono-diet: "to eat little and of few things is a good guide for health and for simplicity." By building cellars to store root crops, making sauces and juices from fruits, and drying herbs, they extended their supply of vegetables and fruits. All in all, the Nearings were able to provide themselves with about 80% of their food.

Their lives were hardly defined by constant work, though. In fact, they ran the homestead each working a mere four hours a day. The remainder of the day was reserved for reading, playing music, perhaps going for a stroll. The Nearings were able to finance their homestead entirely from the production and sale of maple syrup, as well as relying on traditional bartering.

Toward the end of *Living the Good Life*, the Nearings cite as the overarching reason for their move a desire to align their personal theories on how best to carry out an ethical and purposeful life with the actual practice of living. In New York, for them there was a disjuncture between the theory and the practice of living.

BACK TO BASICS: Helen and Scott Nearing (pictured at left) reconciled theory and practice by leaving New York City in 1932 to create what they called a "self-sufficient household economy" on a 65-acre farm in a small Vermont village.

In this sense, what the Nearings set out to do in a mountain valley in Vermont was nothing new per se; rather, they fit neatly along a continuum of historical figures who have abandoned a perceived corrupt "civilization" in exchange for a simpler, rural existence, including Henry David Thoreau and Christopher McCandless (made famous in Jon Krakauer's *Into the Wild*).

Another strain in American history, originating in the Jeffersonian era, is a disdain for the perceived corrupting tendencies of urban centers and the idealization of a rural, agricultural existence. Mark Twain once wryly quipped, "Civilization is a limitless multiplication of unnecessary necessaries."

The conversation contained in *Living the Good Life* can also be found in contemporary works such as *The Omnivore's Dilemma* by Michael Pollan and *Animal, Vegetable, Miracle* by Barbara Kingsolver.

At the root of this centuries-old conversation is the predicament of how best to lead an ethical, purposeful life within an economic system that some of us consider unworkable in its present state. Do we attempt to carve out sustainable spaces within the system, reforming incrementally from within, or do we remove ourselves entirely and practice our theories on living the good life elsewhere?

The Nearings would answer, "Under any and all conditions one is responsible for living as well as possible within the complex of circumstances which constitutes the day-to-day environment." Perhaps this advice is a good starting point for all of us to consider what might constitute our own Good Life, and then to act accordingly.

Matthew Bachler is a native of Vermont. He lives in Los Angeles, where he is the assistant manager of the Hollywood Farmers' Market and an advocate for urban farming.

TO THE ORES: Russ George (in Hawaiian shirt) looks over the mixing barrel as his team deploys iron oxide in the North Pacific, in 2002.

# Plankton Evangelist

The eccentric saga of Russ George. By Charles Platt

Back in 2002, eco-maverick Russ George buttonholed veteran rock star Neil Young and asked an unusual favor. George was living part-time on a sailboat on a dock in the San Francisco Bay Area, and had learned that Young's 90-year-old Baltic schooner was moored just four slips away. With the in-your-face enthusiasm that is his trademark, George brazenly asked to hitch a ride on Young's boat so that he could perform an experiment to revitalize the ocean.

The idea was simple enough: scatter some finely powdered hematite, or red iron oxide, which phytoplankton require as a kind of catalyst when they grow. Phytoplankton are the tiniest sea plants, at the bottom of the oceanic food chain. Krill eat them, and fish eat krill, and penguins and polar bears eat fish.

George liked the idea of sustaining ocean wildlife, but this wasn't his only motive. When phytoplankton grow, they absorb about 100,000 atoms of carbon for every atom of iron they consume. Do the math, and you find that just 1 ton of iron could fix 367,000 tons of $CO_2$. Here, then, was a bargain-basement plan for carbon sequestration. George's ultimate goal was nothing less than to reverse the greenhouse effect.

Young didn't say yes but he didn't say no. The next day, however, George received a visit from the captain of Young's vessel, who told him that he could simply borrow the boat and crew for a month and a half.

Initially George had a bit of trouble rounding up some affordable hematite dust. He needed particles about 0.5 micron in diameter, to minimize their sink rate and keep them suspended in the upper layer of the ocean where phytoplankton live. A mineral processing company offered to custom-mill some iron ore for $10,000 a ton, but when he pleaded

Photograph by Tim Smith

poverty they took pity on him and told him there was a simpler option. He could buy red iron oxide powder as a paint pigment, off the shelf for $700 a ton, from a supplier in Oakland.

So George drove his Ford Taurus over to Oakland with a couple of plastic barrels, filled them with the compound, picked up a couple of bilge pumps and some other necessary items, and in a modern rewrite of a Robert Louis Stevenson adventure, set sail to Hawaii on his personal mission to save the planet.

Four hundred miles east of the Big Island, he mixed the paint pigment with water, pumped it over the side, and watched as it formed a broad, red swath behind the boat. Later he checked data from NASA's SeaWiFS satellite, which has a chlorophyll camera mounted on it, taking pictures in the 552-nanometer spectral line. He felt pretty sure that his mission had been a success.

## A Garage Experimenter

The son of a nuclear chemist with the Atomic Energy Commission, young Darcy Russell George grew up in a house full of science. "Every room in the house had physics and chemistry journals lying around," he recalls. "I used to wander through the desert with my father, scrounging uranium ore."

Today George, 59, lives near Half Moon Bay, Calif., and describes himself self-deprecatingly as a "garage experimenter." Yet he received a Ph.D. equivalency from the Canadian government after he abandoned his U.S. education to avoid the Vietnam draft, and he has a colorful history of talking major laboratories such as SRI International into letting him use their facilities. He even secured an invitation to appear before the House Select Committee on Energy Independence and Global Warming, in July 2007, to make a presentation on the phytoplankton seeding concept.

Over the years, George has started several "cold fusion" and carbon offset companies, and critics say he can exaggerate the scientific basis and potential of his ventures. George concedes he's a risk-taker but says his critics have their own agendas. Yes, he says, plankton seeding needs more research; that's exactly what he's trying to do.

The seeding idea originated with a Moss Landing, Calif., oceanographer named John Martin. In 1986, Martin came up with a new and simple proposal to explain why phytoplankton populations are so sparse in the Antarctic (Southern) Ocean, the equatorial Pacific, and the Gulf of Alaska. He thought it

Phytoplankton absorb 100,000 carbon atoms for every iron atom they consume. Do the math: just 1 ton of iron can fix 367,000 tons of $CO_2$.

could be simply a matter of iron deficiency. Closer to shore, especially where the climate is dry, plankton thrive on iron-laden dust that blows out and settles into the sea.

After performing a few experiments, Martin confirmed his hypothesis and published a paper in *Nature*. He then calculated that phytoplankton could be so amazingly effective at taking $CO_2$ out of the atmosphere that a few hundred thousands of tons of iron could undo global warming. "Give me half a tanker of iron and I will give you an ice age," he joked to colleagues at a conference.

The quip lived on to haunt Martin. To many environmentalists it was no laughing matter. From their point of view, the idea was appalling.

First, it would entail tampering with Nature. The United Nations' Intergovernmental Panel on Climate Change has disparaged the whole concept of "geo-engineering," stating that it is "largely speculative and unproven, and with the risk of unknown side effects."

Second, it would be treating the symptoms, not the cause. "Climate change should be tackled by reducing emissions, not by altering ocean ecosystems," complained biologist Dr. Paul Johnston, head of Greenpeace International's Science Unit. In other words, solving the problem of global warming is acceptable only by reforming bad practices, not by fixing things so that industry can continue to do business as usual.

Russ George offers three points in rebuttal.

First, to reduce $CO_2$ concentrations simply by scaling back industrial output will take decades, even if it is economically and politically feasible, which is open to doubt.

Second, carbon sequestration by other methods would be hugely expensive and grossly inefficient compared with the phytoplankton scheme.

Third, there is another, more urgent problem. Satellite assessments indicate that ocean chlorophyll has diminished by 12% in the Southern Ocean, 17%

in the North Atlantic, 26% in the North Pacific, and 50% in subtropical oceans since measurements began in the 1980s. This translates as a loss of 1% per year of "ocean forests," with a major negative impact on fish populations. The obvious way to reverse this trend is by encouraging phytoplankton to grow.

## From the Galapagos to the Canary Islands

After his improvised experiment on Neil Young's boat, George started looking for ways to fund a larger-scale initiative, and saw nothing wrong with the profit motive as a way to move things faster. Carbon offsets sell for more than $10 a ton. What if he could induce phytoplankton to sequester a million tons? He made this pitch to a venture capitalist, who promptly wrote a check for $50,000. "He said he thought I needed some gas money," George recalls.

The VC pledged to raise the rest of the funding for an iron dust dumping mission if George would do the work. He founded a company that he named Planktos Corp., and bought a research ship for $700,000, ready to fulfill Plan A. "We retained a crew, and aimed to sail to the Galapagos Islands," he says. The Galapagos were an ideal location, being surrounded by a large natural plankton bloom in the equatorial Pacific Ocean, an area that's otherwise mostly lifeless. If Planktos could seed the edge of the existing area, the results should easily be visible via satellite.

Unfortunately, George's tendency to talk publicly, enthusiastically, and volubly provoked a massive negative response. It started when he made the mistake of referring to his microscopic dust particles as "nanoparticles."

*Nano* is not a good word in environmental circles. The Canadian environmental group ETC Group took note and launched an all-out offensive, denouncing the plan as a "dangerous experiment." As George tells it, "They sent delegations to the Ecuadorean government, claiming that we would dump American toxic waste which would poison the seas and kill off thousands of unique species. We explained that the Galapagos is a marine oasis precisely because it already has iron in the ocean around it. But the Ecuadorean government blocked us."

Undeterred, he tried Plan B. Near the Canary Islands, off the coast of Africa, 500 million tons of dust blow into the ocean every year. "But there's a window when the African dust doesn't blow,"

George explains. "If you put iron into the water in December or January, you should get a three-month plankton bloom."

He received permission from all the relevant authorities and was ready to sail from Miami when two armed agents from the EPA turned up at George's office. Apparently, ETC and Greenpeace had alerted them. Since the ship wasn't carrying any iron dust at that point, it could not be prevented from leaving harbor. So, the team sailed for Bermuda — where they were stopped and inspected again. "The Spanish press described us as a toxic waste pirate vessel," George recalls.

Ironically, George had impeccably green credentials, having been an early Greenpeace activist. "I sailed on *Rainbow Warrior* to save the whales," he says. "And for the voyage to the Canary Islands, I'd hired Peter Willcox as the captain." Willcox had captained *Rainbow Warrior* and is a hero among environmentalists, even meriting his own Wikipedia entry.

It made no difference; Greenpeace was still opposed to plankton seeding, and George believes that the "dark green" groups' motives were not entirely pure. "They were making millions in donations from their campaign against Planktos," he says. When he reached the Canary Islands, the Spanish government buckled under pressure from environmentalists and refused permission. The mission was aborted, and George lost his investors.

## Don't Call It Carbon Sequestration

In the meantime, the Alfred Wegener Institute for Polar and Marine Research (AWI) in Bremerhaven, Germany, had been pursing its own plankton plan on an academic, nonprofit basis, using its research vessel, the *Polarstern*. With deeper pockets, better credentials, and less in-your-face enthusiasm, AWI emerged victorious from its own series of battles with outraged environmentalists and announced in January 2009 that it had secured all necessary permits to conduct an iron-seeding voyage.

The institute insisted, however, that this had nothing to do with the controversial topic of carbon sequestration. "A large number of reports are circulating on the internet and in the international press claiming that the Alfred Wegener Institute is conducting the experiment to test the geo-engineering option of ocean fertilization as a means to sequester large quantities of carbon oxide from the atmosphere," Dr. Karin Lochte, director of AWI, noted in a prepared statement. "This is definitely

Photography by Adam White (*Weatherbird*), Tim Smith (*Ragland*)

FEEDING PHYTOPLANKTON: (clockwise from top) Planktos Corp.'s research vessel *Weatherbird* in the mid-Atlantic, late fall 2007; the 2002 voyage to Hawaii in Neil Young's schooner *Ragland*, spreading a trail of red iron oxide to stimulate plankton growth; George cleans up after deploying iron on the *Ragland*.

not the case. … We hope that through this experiment we will be able to contribute to a better understanding of ocean biogeochemistry and pelagic ecosystem functioning."

Indeed. The satellite data, however, spoke for itself. After the *Polarstern* released 6 metric tons (6.6 tons) of iron on Jan. 27, satellite images showed a clearly visible plankton bloom on Feb. 14.

George was delighted, issuing his own press release proclaiming that the phytoplankton "will produce hundreds of thousands of tonnes of krill and other zooplankton. The next step on that food chain are the baby calves of the Southern Ocean Great Whales, as the new pasture is within their traditional nursery. The food chain formula tells us to expect tens of thousands of tonnes of whales being nourished from this wonderful gesture."

By this time he had abandoned Planktos Corp. and started a new venture named Planktos Science. "We're now a private company," he explains, "no longer required by the SEC to be transparent, and we're going to keep our mouths shut, because we know that there are groups that will spend millions of dollars to kill us."

He remains convinced that money is a major issue motivating his adversaries. "The first Kyoto target is to solve 10 percent of the $CO_2$ problem," he explains, "which many people believe will cost around $400 billion. Now, suppose we come along and say we may solve 50 percent of the problem for $4 billion. If you are in a company expecting to get a slice of the $400 billion, you're going to be pissed."

In March, AWI reported that its Southern Ocean experiment captured little carbon, because a previous plankton bloom had depleted key nutrients. Other regions, George says, would be more suitable.

In any case, like AWI, George is not citing carbon sequestration as his primary goal anymore. "My best traction right now is to deal with governments who are getting desperate about their fisheries," he says. Fish, after all, ultimately feed on phytoplankton.

Whether this strategy will keep him under the environmental radar remains to be seen. If he fails to keep a low enough profile, and his diplomatic skills prove to be insufficient or ineffective, George may yet have to try Plan C in his mission to save the planet: persuade Neil Young to buy a bigger boat.

---

Charles Platt is currently writing an introductory guide to electronics for Make: Books.

# Maker

**THE FUTURE STARTS HERE:** Big things are coming from America's garages. This one belongs to our own Mr. Jalopy.

# What's in Your Garage?

We're betting on solutions to big problems coming from innovative makers working in their basements, garages, and workshops. By Dale Dougherty

In an October 2008 presidential debate, moderator Tom Brokaw asked whether serious challenges such as climate change could be met by big Manhattan-style projects like the one that developed the atom bomb, or by people working in 100,000 garages, which is how Silicon Valley started.

Both candidates wandered off topic, but I really wanted them to answer the question. As a matter of national policy, do we think change will come from ideas developed by the "best and brightest" minds? Or will it come from grassroots innovation, widely distributed and wildly varied?

When I heard the question, I replied on our makezine.com blog: "A lot is going to depend on people working in 100,000 or more garages,

probably with little funding or support. ... Our goal is to find some of those industrious, ingenious makers at work in garages everywhere."

Like fabled garage bands, Silicon Valley startups, and Mister Jalopy's bike repair workshop (above photo), maker-led businesses have been started in all kinds of unusual spaces. Here are a few examples:

**Limor Fried** moved into a live/work space near Wall Street that she got for the right price because young investment bankers were fleeing the city. She's come up with numerous original designs, including the Proto Shield for Arduino and MiniPOV kit, and she set up Adafruit Industries (adafruit.com) to bring those designs to market. A pioneer, along with

Photograph by Dave Bullock

Phil Torrone, in the open source hardware movement, Fried's most recent project is the award-winning Tweet-a-Watt (*see page 112*), which allows you to monitor your electricity usage via Twitter.

**Nathan Seidle** started SparkFun Electronics (sparkfun.com) in his apartment, as he finished up an EE degree from the University of Colorado at Boulder. Initially, he tried making products based on surface-mount devices (SMD) without buying an expensive reflow oven. Kitchen toaster ovens toasted the components while melting the solder paste. Hot plates would heat the bottom of the board only.

"At one point, we had four people building boards and reflowing on the hot plate," he says. After three years working with crock pots and hot plates, Seidle decided in 2006 to buy a used reflow oven for $5,000.

Today, his production operation is staffed by 20somethings who have learned about electronics on the job. It's a factory without an assembly line — a mix of machines, tools, computers, and engaged workers learning new things. SparkFun has become the new RadioShack, with 50 people working on two floors of an office park whose previous tenant moved their manufacturing to Alabama and Thailand.

**Kyle Wiens** and **Luke Soules** started iFixit (ifixit.com) out of their dorms at Cal Poly in San Luis Obispo, Calif. That was six years ago. Today they have a self-funded business that sells the parts and tools you need to repair Apple equipment. One of their innovations is creating online repair manuals for free that show you how to make the repairs.

"Our biggest source of customer referrals is from Apple employees, particularly folks at the Genius Bar," Wiens says. They refer customers who complain when Apple won't help them fix an out-of-warranty product. (Apple: "Just buy a new one!")

iFixit will also buy your old Mac and harvest the reusable parts to resell. (There are about 30 such parts in a MacBook.) If it's starting to sound like an auto parts franchise, well, Wiens and Soules have been thinking about someday doing for cars what they do for computers and handhelds today.

**Derek Elley** is chief strategy officer for Ponoko (ponoko.com), a company based in New Zealand that allows you to design and manufacture almost any 2D design by having it custom-cut on a ShopBot or laser cutter. Recently, Elley told me Ponoko was opening factories in the United States, starting in Oakland,

Calif. I asked him where and he said at "Because We Can." I was surprised, but I shouldn't have been.

Oakland-based Because We Can (becausewecan.org) was started by **Jillian Northrup** and **Jeffrey McGrew** out of the small converted barn that was their house. The couple was among the first makers to master a ShopBot, bringing it to the first Maker Faire to show off their coffee tables and lap desks. They also tackle larger projects such as fantastic office interiors, one even inspired by Jules Verne. They recently moved to a commercial space, which McGrew says has a roll-up door and gets cold like a garage, "but it's not really a garage."

As ShopBot (shopbottools.com) founder **Ted Hall** thought about 100,000 garages from his Durham, N.C., headquarters, he saw it as a big idea about digital fabrication. Was it possible to build a network of 100,000 garages, sharing similar sets of tools and capabilities? Could each garage become part of a highly flexible and distributed manufacturing network? Could work be distributed to these garages via the internet, so that individuals and companies who create or design things could have them made locally?

Working with **Bill Young**, also of ShopBot, Hall has set up 100Kgarages.com to spearhead this effort, creating new jobs and opportunities for makers. "New digital fabrication tools," Hall writes on the site, "allow us to efficiently make things that would have just a few years ago been at best difficult, and most likely impossible, using traditional tools and methods."

Hall began talking to Ponoko about providing the software infrastructure for the new garage network. So watch for "Bill and Ted's" next excellent adventure, which might take American manufacturing in a completely new direction.

On the internet, nobody knows that your factory is really a garage.

*This summer, MAKE will open Makers Market (makersmarket.com), an online marketplace, much like a farmers market, where people who make things in all kinds of spaces can find more buyers. (Read more about it on page 15.) Please check it out and let us know what you think.*

---

Dale Dougherty is the founding editor and publisher of MAKE.

# Putting His Tools Into Circulation

Dustin Zuckerman's tool library. By Dale Dougherty

**R**eserved is a word with multiple meanings for a librarian, even one who doesn't lend books. Soft-spoken and reserved, Dustin Zuckerman works the periodicals desk at the Santa Rosa Junior College Library by day. On his own time, he runs the Santa Rosa Tool Library out of his modest apartment.

Patrons of the tool library can go online to reserve tools for a weekend project and then drop by Zuckerman's apartment in downtown Santa Rosa, Calif., to pick them up. On a summer Saturday, about 15 people will come to borrow tools such as disc sanders, jackhammers, drills, pickaxes, pressure washers, and 100-foot tape measures.

Zuckerman, 38, started the tool-lending library in 2008 for a practical reasons — he himself had needed tools he couldn't afford to buy or rent for a gardening project. But he also saw the lending library as a way to help others. "I got to thinking that I wanted to have something to show for myself," he says. "So I started this project."

When he explained the idea to friends, they surprised him with a $200 gift card from Sears, where he bought a set of plumbing tools to seed his library. He made several visits to his family's pawnshop in Los Angeles, where he filled up his van with donations such as a rotary hammer drill.

Once he got the library going, patrons donated tools they didn't need or didn't use much. Now he has more than 700 tools of 300 types; those most

Photography by Sam Murphy

frequently borrowed are stored in a closet in his apartment. Others are in a nearby garage.

Zuckerman set up his lending library by purchasing circulation software used by small libraries, and created a website at borrowtools.org. "It's good to think like a librarian in setting up a tool-sharing service," he says.

Users register online with their driver's license or ID to borrow a tool. There are no fees for borrowing, but you have to sign a borrowing agreement. While there are late fees, Zuckerman says, "We rarely have overdue tools." Only twice have tools been broken, and each time the borrower showed up at his doorstep with a new replacement.

More women than men use his lending library, including a handful who have little experience with power tools. He spends time giving each person a tutorial in how to operate the tools, and provides goggles and earplugs, plus a hard hat, if needed. "I'm not an expert on tools," he says. "I just try to learn the basics so I can pass it on."

One woman borrowed a kit to change the oil in her car, and Zuckerman says that when she returned it, she had a big smile on her face and told him how much she enjoyed being able to do the oil change herself. This is the big reward for Zuckerman: his borrowers are particularly grateful.

Zuckerman is a middleman, making it more comfortable for people to borrow. Borrowing a tool from a friend or neighbor seems to create an uneasiness around obligations. He sees himself as a librarian abiding by a clear set of rules for lending things that ultimately belong to the community. His job is to facilitate sharing.

One day Zuckerman hopes to have a small storefront and warehouse, along with a mobile unit to drop off or pick up tools. Such ambitions may require fundraising by his new board of directors, or user fees. For now, it's a one-person, part-time operation.

Tool lending libraries can be found around the country (look on Wikipedia for a list) including those affiliated with public libraries in Berkeley and Oakland. If you're thinking of starting a tool lending library in your community, Zuckerman recommends working with a librarian if you're not one yourself. "Start small and test it out with a few people," he advises reservedly.

---

Dale Dougherty is editor and publisher of MAKE.

# Building a CNC Mill

There's a lot of well-deserved excitement surrounding the RepRap 3D printer, and much of it focuses on the RepRap's ability to make its own parts. The RepRap's fabrication technique is *additive* — it uses a plastic extruder to "print" a plastic model, one layer at a time.

This contrasts with *subtractive* fabrication techniques, which start with a solid block of material and use a cutter to remove the excess.

Subtractive fabrication is far more common than additive, especially when working with metal and wood. Lathes, mills, saws, and drills are all subtractive tools. A CNC milling machine or router is the subtractive equivalent to the RepRap 3D printer.

For the hobbyist, milling is inferior to printing in numerous ways. The method inherently causes waste, and without any sort of dust control, that waste gets flung throughout the room.

Milling is more hazardous: while it's possible that a plastic extruder might overheat and catch fire, I've already had a (minor) fire with my CNC router, and there's the added danger of a blade, spinning at 20,000rpm, sending bits of itself, or even your workpiece, flying at you.

A mill or router is necessarily larger, heavier, and consequently more expensive and more difficult to move. It requires a positioning system that can maintain accuracy when encountering resistance, and motors powerful enough to drive it.

Software preparation is also more complex for milling. After drawing the object you wish to make in a CAD or 3D modeling program, it's necessary to generate tool paths with CAM (computer-aided manufacturing) software. This involves specifying the dimensions and location of the stock material, the dimensions and characteristics of the end mill (cutter), and speeds for the axes and spindle.

The tools to do all this tend to be complex and a bit daunting for the first-time user. From the user's perspective, CNC milling is much more complex than printing.

CNC milling does, however, have a significant advantage over 3D printing: the technology is mature. The RepRap is improving at a tremendous rate, but there's still a lot of tinkering and experimentation involved in getting a good print. If 3D printing technology fascinates you, the RepRap is a great project to get involved in, but if your interest is in making things, CNC milling is the better option at this time.

Going with CNC milling, however, does not mean you have to give up on self-replication or on making your own machine.

---

CNC milling has a significant advantage over 3D printing: the technology is mature.

If 3D printing technology fascinates you, the RepRap is a great project to get involved in — but if you want to make things, CNC milling is the better option at this time.

---

Patrick Hood-Daniel's Expandable CNC Kit (buildyourcnc.com) is a CNC router capable of making all of its custom parts. The router's frame is built of custom-cut MDF (medium density fiberboard). Everything else is standard hardware. The aluminum angle, bolts, and screws are available from any hardware store.

The lead screws and anti-backlash nuts will probably have to be mail-ordered from McMaster-Carr and dumpstercnc.com, but you can get by with lesser hardware store parts in a pinch. The stepper motors and stepper drivers are completely generic and available from countless sources. The spindle is an ordinary wood router; I use a Porter-Cable 892.

As with the RepRap, the trick with Hood-Daniel's CNC router is getting a seed unit. Fortunately, any CNC router that can handle a 2'×4' sheet of MDF is capable of making the parts. Start looking around the forums at makezine.com or *The Home*

DIY CNC: Patrick Hood-Daniel's Expandable CNC Kit is a CNC router that can make all of its own custom parts.

*Shop Machinist* (bbs.homeshopmachinist.net), and you're likely to find somebody local with the necessary equipment.

To get the files, fill out the contact form at buildyourcnc.com and Hood-Daniel will email you the CAD and CAM files for the parts. Videos on his website explain how to put everything together. The files are licensed under the Creative Commons Attribution-Noncommercial license, which essentially means you can make machines for yourself and for your friends, but you can't sell them commercially.

The CAD files are in CamBam format. CamBam is proprietary, but there's a free version available and it can export to DXF. Most good CAM software is very expensive, though, so if you don't already have a favorite, stick with CamBam. It can do both CAD and CAM, and even the free version is tremendously full-featured.

If you can't find anybody to cut the parts for you, Hood-Daniel sells a kit with the basic hardware for $1,100.

The first thing you'll want to do, once you get it running, is make a set of spare parts. Next, start making replicas for your friends. While the MDF frame may look a bit light if you're used to professional equipment, I've had no problem cutting wood with it. The design isn't rigid enough to cut metal, but you could always cut out a new tool holder and mount a plasma cutter.

The platform could also support a RepRap plastic extruder — so when 3D printing *does* take off, you'll already have a CNC platform ready with a 2'×4' printing area.

Tom Owad is the owner of Schnitz Technology, a Macintosh consultancy in York, Pa. He spends his days tinkering and learning, and is the owner and webmaster of applefritter.com.

# How to Analyze Scientific Images

Digital cameras are among the most important instruments in my science tool kit. They have provided thousands of images of twilight glows, solar aureoles, clouds, tree rings, insects, vegetation, bacteria and mold colonies, and much more.

Most images speak for themselves and need no more processing than that applied by your eyes and brain. But what if you want to extract data from images? Analytical studies of photographs require more than simply describing, say, the color of a leaf or the brightness of a sunset — they require numbers.

Prior to the digital era, scientists relied on instruments called *densitometers* to extract data from photographs. Transmission densitometers convert the degrees of darkness (or density) of points on photographic film into a representative voltage or signal. One side of the film is illuminated by a light source, and the light that leaks through is detected by a photosensor on the opposite side. Photographic prints are digitized by placing both the light source and the detector on the same side of the image.

While densitometry is still used to extract data from photographic film and prints, these can now be easily digitized by what amounts to a new kind of densitometer: the flatbed scanner. And the data in digital images is already present in a form that can be easily analyzed by image processing software.

## ImageJ: Image-Processing Software that's Powerful, Platform-Independent, and Free

Processing digital images to extract their data once required powerful, expensive software. Its high price prevented many students and amateur scientists from analyzing their images.

This changed in 1997 with the introduction of ImageJ, a public-domain image analysis program developed by Wayne Rasband at the National Institutes of Health (NIH). During its first decade, ImageJ became a powerful, platform-independent image analysis package that can be run on Linux, Macintosh, and Windows machines. I've run it on a variety of Windows machines, including a tiny Acer Aspire One running Win XP with only 1GB of RAM (and a 160GB hard drive).

ImageJ requires no license, and the program and its Java source code are freely available at the ImageJ homepage: rsbweb.nih.gov/ij/index.html.

## Running ImageJ

After you download the version of ImageJ designed for your operating system, go ahead and select and run the program. A rectangular command bar will appear at the top of your computer screen (Figure A). This tiny but powerful startup menu is a toolbox that includes a set of 8 text selections over a row of 19 icons that point to menu options known as macros, some of which can be accessed immediately while others lead to still more macros.

ImageJ's startup menu floats over the upper edge of whatever program was running when you clicked on it. You can use your mouse to drag it anywhere on your screen. If the toolbox disappears, simply click on the ImageJ icon on your task bar to place it back on your screen.

## Doing Science with ImageJ

Let's get started with ImageJ by analyzing a photograph taken with a digital camera. The image in Figure B reveals a glow around the sun known as the *solar aureole*. The sun itself is blocked by the occluder device I described in my previous column (*see MAKE, Volume 17, page 48*).

First, select File ⇒ Open and choose an image, in this case the JPEG in Figure B. After the image is displayed on the screen, maximize it. You're now ready to analyze the image.

This image was taken on a clear day at solar noon, and it can tell us much about the presence of aerosols in the sky, such as smoke, dust, and haze. You can quickly see how — simply place the cursor over various parts of the image and watch the 5 numbers that appear below the row of toolbar icons.

The first 2 numbers indicate the x–y coordinates of the position of the cursor on the image. The next 3 numbers indicate the intensity of the red, green, and blue (RGB) wavelengths of light at that position, on a scale from 0 to 255. Placing the cursor over the

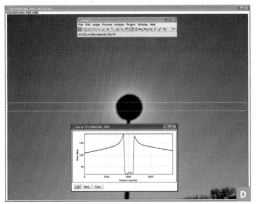

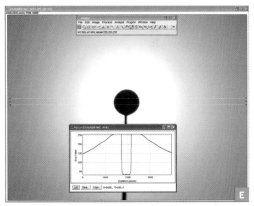

Fig. A: The ImageJ menu window. Fig. B: Solar aureole image ready for processing in ImageJ.
Fig. C: Selecting a region to be plotted. Fig. D: Intensity plot of the region selected in Figure C.
Fig. E: Intensity plot of a solar aureole caused by African dust over Texas.

black occluder will cause each of the RGB numbers to fall below 20. Move the cursor over the nearby sky, and the blue number will increase much more than the red and green. The blue number will be highest when the cursor is moved all the way to the edge of the image.

Now it's time to let the program do all this for you, by drawing a graph of the intensity of blue light across the center of the image. First, make sure the ImageJ menu bar is displayed. Then click-drag with your mouse to draw a narrow rectangle across the entire image, centered over the black occluder. The rectangle will be outlined in yellow, and you can use your mouse to alter its size or nudge the entire rectangle into a different position. This process is called making a selection (Figure C).

Now you're ready to analyze the intensity of

sunlight within your selection. Select Analyze ⇒ Plot Profile. A graph depicting the intensity of light across the rectangle you selected will almost instantly appear (Figure D).

The solar aureole plotted in Figure D was for a clear sky. Figure E shows the plot for a day when a considerable amount of Saharan dust had blown from Africa to Texas. The images are clearly very different and so are the plots.

The plots allow you to list the digitized values, which you can copy and paste into a spreadsheet like Excel or the free, open source Open Office Calc (openoffice.org). You can then make a chart like the one in Figure F (following page), which superimposes the data in Figures D and E to clearly see the huge difference between the clear and dusty skies.

Photography by Forrest M. Mims III

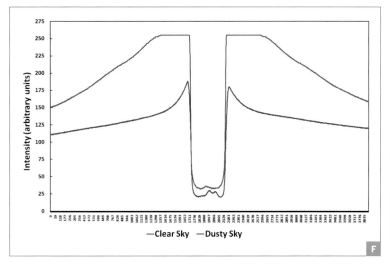

Fig. F: Superimposed intensity plots of solar aureoles on a clear day and a dusty day. Figs. G and H: Two 3D surface plots of solar aureoles, in a clear sky and a dusty sky, respectively.

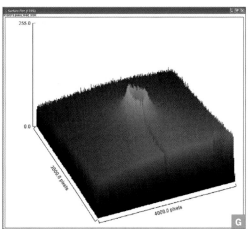

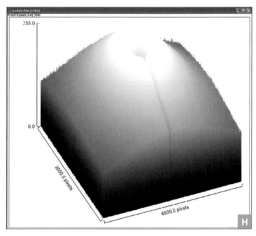

## Going Further

We've barely begun to exploit the capabilities of ImageJ with these simple examples. Want to see the distribution of separate red, green, and blue wavelengths? Select Image ⇒ Color ⇒ Split Channels. The program will quickly present you with 3 grayscale images, for R, G, and B. You can then analyze each image any way you like. For example, Figures G and H show 3D surface plots for the red wavelengths of both the clear and dusty aureoles above.

You can learn much more about ImageJ from the website, but the best way is to select some nice images, download the software, and take the plunge. You'll be amazed by this program's capabilities.

While it's fun to explore what ImageJ can do, you can use it for serious science. For example, image analysis software was the only way I was able to write a scientific paper based on an analysis of solar aureole images made with an old 1998 Fuji with only 1.3 megapixels resolution (F.M. Mims III, Solar aureoles caused by dust, smoke and haze, *Applied Optics* 42, 492-496, 2003).

Of course ImageJ isn't limited to analyzing sky photos. One of my favorite activities is studying tree rings. In the next issue of MAKE, we'll see how ImageJ can transform this hobby into serious science, too.

---

Forrest M. Mims III (forrestmims.org), an amateur scientist and Rolex Award winner, was named by *Discover* magazine as one of the "50 Best Brains in Science."

# ⚡ REMAKE: AMERICA

A global economic crisis isn't something anyone would wish for, but that doesn't mean it's all bad. These challenging times have presented us with a rare chance to try out new ways of doing things. The opportunities for makers are terrific — we can start at home to remake manufacturing, education, food production, transportation, and recreation. Here are some projects to get you started.

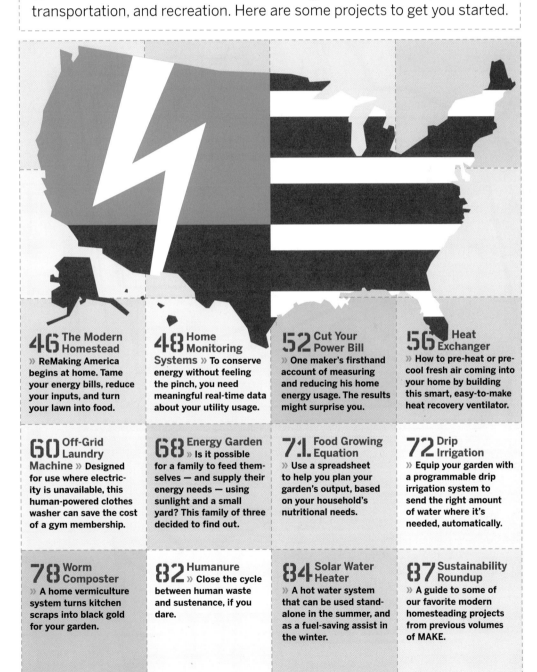

**46 The Modern Homestead**
» ReMaking America begins at home. Tame your energy bills, reduce your inputs, and turn your lawn into food.

**48 Home Monitoring Systems** » To conserve energy without feeling the pinch, you need meaningful real-time data about your utility usage.

**52 Cut Your Power Bill**
» One maker's firsthand account of measuring and reducing his home energy usage. The results might surprise you.

**56 Heat Exchanger**
» How to pre-heat or pre-cool fresh air coming into your home by building this smart, easy-to-make heat recovery ventilator.

**60 Off-Grid Laundry Machine** » Designed for use where electricity is unavailable, this human-powered clothes washer can save the cost of a gym membership.

**68 Energy Garden**
» Is it possible for a family to feed themselves — and supply their energy needs — using sunlight and a small yard? This family of three decided to find out.

**71 Food Growing Equation**
» Use a spreadsheet to help you plan your garden's output, based on your household's nutritional needs.

**72 Drip Irrigation**
» Equip your garden with a programmable drip irrigation system to send the right amount of water where it's needed, automatically.

**78 Worm Composter**
» A home vermiculture system turns kitchen scraps into black gold for your garden.

**82 Humanure**
» Close the cycle between human waste and sustenance, if you dare.

**84 Solar Water Heater**
» A hot water system that can be used stand-alone in the summer, and as a fuel-saving assist in the winter.

**87 Sustainability Roundup**
» A guide to some of our favorite modern homesteading projects from previous volumes of MAKE.

# Energy Production, Use, and Monitoring with Some Geeked-Out Gardening

## A bird's-eye guide to MAKE's modern homesteading projects.

ILLUSTRATION BY DAMIEN SCOGIN

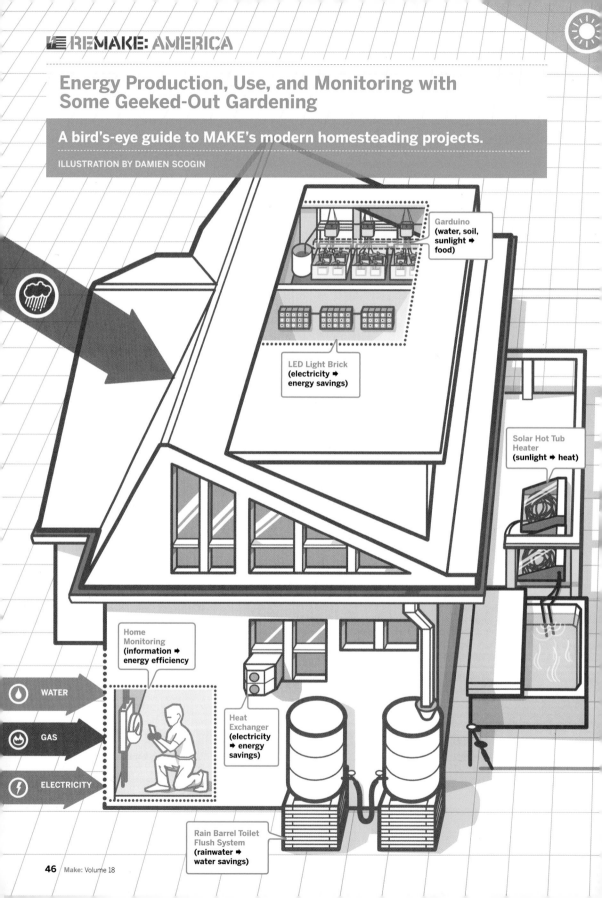

**Garduino**
(water, soil, sunlight ➜ food)

**LED Light Brick**
(electricity ➜ energy savings)

**Solar Hot Tub Heater**
(sunlight ➜ heat)

**Home Monitoring**
(information ➜ energy efficiency

**Heat Exchanger**
(electricity ➜ energy savings)

WATER

GAS

ELECTRICITY

**Rain Barrel Toilet Flush System**
(rainwater ➜ water savings)

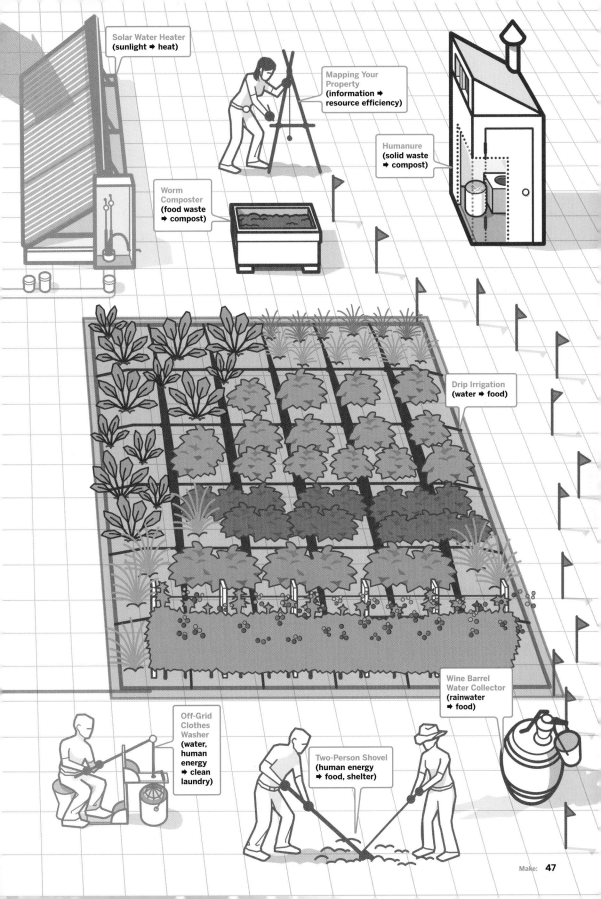

Solar Water Heater
(sunlight ➡ heat)

Mapping Your
Property
(information ➡
resource efficiency)

Humanure
(solid waste
➡ compost)

Worm
Composter
(food waste
➡ compost)

Drip Irrigation
(water ➡ food)

Wine Barrel
Water Collector
(rainwater
➡ food)

Off-Grid
Clothes
Washer
(water,
human
energy
➡ clean
laundry)

Two-Person Shovel
(human energy
➡ food, shelter)

# Home Energy Dashboards

**G**ood science begins at home. I learned this the day the owners of a fancy automated house let me loose inside their place for an hour to check out their energy monitor. I ran amok, letting the jacuzzi fill up, burning toast for no reason, and flipping on and off an expensive plasma slab TV.

The results were intoxicating: a tabletop display in the kitchen showed pretty, Flash-animated bar graphs rising to an alarming crest, tallying an exact count of kilowatt-hours, gallons, and therms wasted, and tripping cute animations to point out that I'd used more resources than Al Gore's mansion during a global-warming fundraiser.

Of course, the whole point of resource monitoring is to save energy. But it's hard to ignore the allure of real-time feedback. Energy's usually shrouded in mystery. Your bill comes once a month, and then who knows what caused a spike or trough in your consumption?

"People need constant reminders of what's going on in their houses, and monitoring your usage can have a big impact on behavior," says Ed Lu, lead engineer of Google's energy monitoring group and a former NASA astronaut. "If I took the speedometer out of your car, you could guess at the speed pretty close, but not that close."

What I learned from my hour of prodigal excess, however, is that instantaneous data is only the beginning. If I were an actual occupant of the house, I would have been looking at the historical trends and the energy display's little advisory messages all along.

It's the richness of the data that makes it useful. After all, Cadillacs sported simple gas mileage displays back in the 1970s, but it was the triumph of the Toyota hybrid user interface that gave us the "Prius effect": a graphic, 30-minute historical trend display that pushes drivers to new MPG goals.

In a household setting, analysis needs to be even more sophisticated. A good energy monitor will let you filter out high-wattage appliances like toasters or hot glue guns that are only on for a short time, and reveal pesky resource hogs like faulty pool pumps, hibernating desktop PCs, always-on stereo receivers, and failing fridges that cause long-term waste.

Says Collin Breakstone, VP of business development for Agilewaves, whose Resource Monitor was

Illustration by Alison Kendall

designed by former NASA engineers: "Because of our background in mission-critical data acquisition systems, we know how to handle massive amounts of data and display it in a meaningful, actionable way."

We'll need it, especially as home energy pricing becomes more complex. Utilities around the country are starting to offer residential customers the option of a *time-of-use rate*, in which power becomes less expensive after peak hours. There are also *tiered-rate* plans, in which you pay less per kilowatt-hour if you bring down your overall consumption. Utilities eventually want to move to *real-time pricing*, which will fluctuate directly with the market.

To sort through the options, energy monitoring systems like those from startup Greenbox send personalized messages to your web browser, such as: "Your base load increased by 60 watts yesterday. You would save $162 a year by switching electricity plans."

Such straightforward advice is a bonus for homeowners attuned to environmental concerns. And people like to save money. But will the money saved on a typical $100 bill (maybe 15 bucks a month) motivate everyone?

"No one really knows how much people would be willing to change their behavior," admits Michael Murray, CEO of monitor maker Lucid Design Group. "It's not going to be the early adopters who will help us get out of this mess; you have to engage the rest of the population."

To reach the wasteful masses, Lucid created an energy interface that pits users against one another — and buildings against one another — in a game to see who can expend the least wall juice.

"Competition is a powerful way to motivate people," says Murray. "What some people really want is bragging rights. It's about being number one, not necessarily all about saving the planet." Lucid currently installs most of its systems in university dorms, where it sees very good participation in its intramural energy contests.

Meanwhile, other companies are contemplating the competitive drive too. Google's PowerMeter system needs widespread installation of smart electric meters to gain ground, but then users will be able to share their data in many different ways.

"I'm sure we'll see throw-downs of one city versus another city," says Google's Lu. "How about Stanford versus Berkeley?"

After experiencing the immediacy of real-time data firsthand, I wanted a dashboard for my own house. And if I couldn't afford it, I would rig one

## "People need constant reminders of what's going on in their houses, and monitoring your usage can have a big impact on behavior."

myself. Here's a representative sample of various options available:

### Agilewaves Resource Monitor
A complete system for homes and commercial buildings. Comes with electricity, water, and gas hardware from industrial suppliers, and a gateway that translates various bus signals from sensors into standard IP protocol. An on-site server (about the size of a Mac mini) or remote host captures and analyzes signals, then displays the data on cellphones and desktops, and automatically instructs subsystems to open motorized windows or fire up heating systems (Figure A, following page).
**Availability: Now**
**Cost: From $7,500**

### Lucid Design Group Building Dashboard
Monitors mainly electricity, and the company works mainly with college campuses. The Building Dashboard Starter product comes with an amp sensor that requires installation by an electrician. The sophisticated interface shows real-time consumption, historical trends, translation into novel and surprising metaphors, and competition statistics between dorm buildings (Figure B).
**Availability: Now for big customers;**
**pilot program for residential customers**
**Cost: From $9,950**

### Google PowerMeter
A pretty face for data that's been harvested and logged by your utility's smart meter. For consumers outside smart meter installation areas, Google software may be compatible with upcoming prepackaged home electricity sensors. PowerMeter will also be open and hackable for anyone else. "We're hoping that the DIY community creates new uses of their own," says Lu (Figure C).
**Availability: Mid-2009 for limited U.S.**
**utility customers**
**Cost: Free**

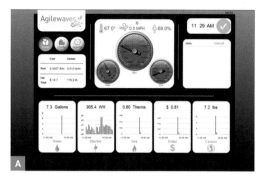

A

B

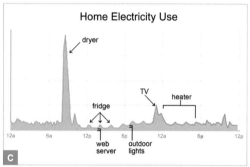

### Home Electricity Use

C

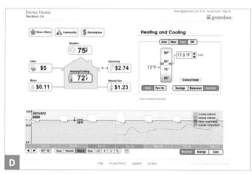

D

## Greenbox

A clean, web-based interface written with Flash software (company founders are the inventors of Flash). Excellent advisory data on payback performance of solar electric and hot water systems. Wireless thermostats connected through ZigBee or low-power wi-fi respond to web requests you make (Figure D).
**Availability: Limited utility customers**
**Cost: Service is free; smart meters underwritten by local utility, or purchased by DIYers**

## Tendril Residential Energy Ecosystem

A flock of mesh-networked gadgets that talk to each other and report to the Tendril server, giving you a detailed picture of consumption. The ZigBee-based wireless network includes a thermostat that can respond to user-designated rules and utility price changes, and outlet adapters that report real-time power use by receptacle — giving a clearer picture of what or who is using the most watts than single-point smart meters could (Figure E).
**Availability: Limited utility customers**
**Cost: Basic setup underwritten by local utility, outlet adapters extra**

## Do-It-Yourself Options

For companies and individuals alike, hardware is the biggest hurdle to resource monitoring. Most usage sensors are expensive industrial models that output a variety of exotic data bus protocols. For instance, the cheapest way to measure water or gas flow is a $300 in-line meter that requires professional installation and custom programming. Doppler sensors that clamp on the outside of pipes and gauge flow using ultrasound are available, but still spendy (over $1,000). Electricity sensors are the least expensive and the easiest to install.
**Availability: Now**
**Cost: From $80**

» For instance, to hook up **Energy, Inc.'s The Energy Detective** ($144), you clip an electromagnetic sensor to the incoming power lines inside your service panel. You can then view usage on its wireless tabletop display, or through the optional PC software or third-party software.

» Another gadget, **Blue Line Innovations' Power-Cost Monitor** ($119) involves strapping an optical sensor to the outside of your electricity meter and viewing the data wirelessly on a desktop LCD. (Blue Line doesn't currently offer a method for crunching data with your own software.)

» **Energy Optimizers Limited's Plogg electrical outlet sensors** are equipped with either Bluetooth

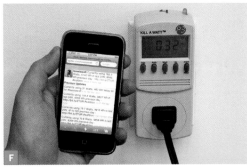

E

F

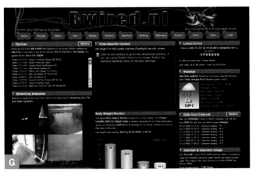

G

($95) or ZigBee ($80) to report back to your custom-coded database wirelessly.

» Some of the most inspired systems come from individual makers. For instance, the thrifty **Tweet-a-Watt system** (see page 112) combines the common Kill A Watt outlet sensor ($25) with XBee wireless communication modules ($23). Limor Fried originally created the network to collect power usage data from around her apartment in real time (Figure F).

The homebrew **Bwired system** from Dutch homeowner Pieter Knuvers (bwired.nl) harvests power, water, and gas data from dozens of off-the-shelf and self-made electrical sensors. Knuvers' system registers every time his family uses the toilet, the bathroom scale, the refrigerator door, and the front door; it also adds up the precise energy used by every appliance in the house and organizes the data neatly online in bar graphs.

While I personally don't need the kind of deep data mining he's getting by cross-tabbing toilets, scales, and fridges, it's nice to see that someone's on the forefront of this new field of domestic research (Figure G).

Bob Parks (distractedmotorist.com) is a frequent contributor to MAKE and works as an editorial consultant for green demonstration homes in Chicago and Los Angeles.

## BE A METER READER

I was impressed with how easy it was to get started monitoring our energy use with the **Powerkuff home power monitor system ($100)**. This device delivers on its promise to bring your electricity meter inside, where you can keep your eye on your ongoing usage.

To install the system, you simply wrap a sensor cuff around your incoming electrical cable or conduit where it feeds your utility meter, no tools required.

The sensor communicates wirelessly to a separate monitor anywhere in your home. Stick in three AAA batteries and the monitor becomes mobile, so you can walk around your house turning things on and off to instantly see the impact on your electricity usage.

It was fun to run around switching things on and off and watching the results. I found out that our home draws about 400 watts during the daytime without any major appliances on. Lamps and small devices like laptops won't show up on the Powerkuff, but I was very interested to notice that while our kitchen baseboard heater bumps the load up 100 watts, our tiny bathroom heater requires four times that much energy. Turning on the electric clothes dryer made the biggest dent, requiring an entire kilowatt — watching our usage jump from 0.4kW to 1.4kW sure made me think a little more about using the clothesline.

This basic electricity monitoring system doesn't offer a lot of bells and whistles or pretty pictures, but it does have a simple data-logging feature so you can track your energy usage beyond the monitor's real-time LCD display. Just download the free software from powerkuff.com and plug the monitor into a PC's USB port to access the data.

The Powerkuff is a good entry point into monitoring electricity use, and it's so easy that it makes a great project for kids. Mine were fascinated by measuring the different amounts of energy our appliances used. Powerkuff also sells an educational kit that's appropriate for classroom use. —*Bruce Stewart*

Bruce Stewart has covered topics ranging from telecom to solar coffee roasters and Lego harpsichords.

Photography by Limor Fried and Phil Torrone (F) and Ed Troxell (below right)

# Power Down with DMAIC

If your electric company has tiered rates, you'll pay a reasonable price for a little electricity, a higher price for more — and a vastly higher price if you use too much.

To reduce my bill, I've started carefully measuring my household electricity usage, and trying different things to reduce it. I'm using the approach that's taught to engineers, called the DMAIC process:

Define — **state the problem to be solved**
Measure — **quantify the problem**
Analyze — **find the root cause of the problem**
Improve — **fix the problem and measure the results**
Control — **make sure the improvements are sustained, and detect new problems that crop up**

Here's how I applied the process to my problem.

## Define.

I don't want to pay the highest rate for electricity.

## Measure.

I need a way to measure energy usage. I wanted to learn, for instance, whether an electric stove is

Illustration by Alison Kendall

A

B

more or less efficient than a microwave. A number of companies have started selling home energy and water usage systems with sophisticated analysis software (*see page 48, "Home Energy Dashboards"*), but I went the less expensive route and bought two different instruments to measure how much electricity my house's appliances and systems were using:

**» The Kill A Watt EZ** (p3international.com) power meter monitors the energy usage of whatever appliance you plug into its standard outlet (Figure A). It measures power on a 110V line up to 1,875W. Using it, I learned that Mr. Coffee uses 850W even after the coffee finishes brewing. Turning it off right away will save me at least $50 a year. I also found some lamps that were still using wasteful incandescent bulbs.

**» The Black and Decker Power Monitor** (blackanddecker.com/energy) is good for measuring the big power hogs and whole-house usage. It isn't as precise as the Kill A Watt, having only 100W resolution, and it only measures down to 300W. But it continuously reads the electric utility meter on the side of your house (Figure B), and wirelessly sends a power consumption measurement to a handheld monitor.

It helped me learn the rhythm of power usage in my house, especially heating, cooling, and refrigeration. To measure a single appliance with it, I turned the appliance on and off, and recorded the change in household power shown. (If other appliances such as the furnace or the refrigerator kick on during the measurement, you have to try again.)

I was disappointed to learn that neither microwaves nor conventional electrical stoves are efficient, and used equal amounts of power to boil 2 cups of water. I asked Kirsten Sanford, Ph.D. (aka Dr. Kiki, creator and host of the radio show and

podcast *This Week in Science* and the TV show *Food Science*) about this. She let me measure her induction stove, which has an element that becomes cold to the touch right after you use it, and it turned out to be about 3 times as efficient as either the electrical stovetop or the microwave.

## Analyze.

What do your measurements tell you about the problem?

» The appliances that use the most juice tend to be expensive, and you may not have the cash to replace them now.

» Some appliances, such as the hair dryer, use a lot of watts but aren't on for long. Others, like my Mr. Coffee, waste power for as long as you'll let them.

» The microwave wins for small jobs and the stove wins for big jobs, but only by a little bit.

» And of course, incandescent lights use way too much power.

I tried a compact fluorescent lamp (CFL) bulb in a ceiling fixture, but it burned out quickly. What was the problem there? A web search found all sorts of complaints about CFLs, but the one that made the most sense was that the electronics in the base tend to get too hot when the bulb operates upside down. That's how the bulb is positioned in my ceiling fixture — inverted.

I measured the temperature of my CFL (see page 55) and found that it ran much hotter upside down — sharply reducing its working life.

## Improve.

The next step was to take action to lower my bill, based on my analysis. Here's what I did.

Photography by Tom Anderson

**C**

1. Using the whole-house energy monitor, I turned energy saving into a game. I wanted to see the meter go down, so I went around turning things off, trying to see how low I could make the number go. I "won" when my house was using less than 300W, which is the lower limit for the meter. By leaving the meter on the kitchen table, it was convenient to "play again" any time.

I quickly became one of those people who go around turning off lights and carefully thinking about what I want before I open the refrigerator. Since I know the power consumption of all my appliances, I know which ones are most important to use wisely.

2. I modified the inverted CFL bulb by cutting the base in two and making a lamp harp to suspend the ballast circuit below the bulb. I ran wires from the base down through the bulb coil, and into the ballast through holes drilled in the side (Figure C, and highlighted box on next page). This solved the overheating problem, but the ballast created an ugly shadow. I'm sure you're familiar with the look of a dead bug in a light fixture — this looked more like a dead rodent. I didn't even bother showing it to my wife. Ideas for improvement are welcomed.

I gave up on the inverted design, for this fixture at least, and switched to an instant-on, high-power-

factor CFL bulb made by U Lighting America. It has a 125°C-rated electrolytic capacitor and it's dimmer-compatible, meaning it also works in lamps that use solid-state switching, such as motion sensors.

3. I changed my always-on Linux server from a desktop machine to a hacked Western Digital NAS drive running Linux on its embedded ARM processor, which uses less power than the desktop server.

4. I found that some appliances used appreciable standby power. I put those on a power strip so that I can turn all of them off. (My new internet radio was the worst offender.)

5. I got an Isolé IDP-3050 occupancy-sensing power strip for my garage, and plugged the main lights into it. I often get distracted and walk away without turning off the garage lights. The power strip senses that I have left and turns them off after an adjustable delay.

6. I also got a Smart Strip LCG3 power strip that automatically switches off outlets based on the draw from its one control outlet. This is ideal for a television, where you'd like to turn off several components at the same time as the TV. The power strip also has unswitched outlets, so that you don't lose time on clocks, for example.

Since I live TV-free, I haven't found a place to use this yet! I was hoping to use it with a computer, but the requirements for the power up and power down sequencing of the computer are too complex for this approach.

## Control.

Control requires ongoing monitoring. The power company reminds me every month how I'm doing! So far, I'm happy to report that my electric bills are lower than they were last year. This is remarkable because I was using CFLs even before I made the changes.

The DMAIC process helped me find a few energy leaks, and cut my usage from 19.9 kilowatt-hours (kWh) per day last year to 16.4kWh/day this year. My baseline rate is 12.6kWh/day, so my reductions put me just under the really high rates, which begin at 130% of baseline.

I'm saving about $28 a month. I'll continue to use DMAIC to see if I can reduce my bill even more.

Tom Anderson is the chief bottle washer at Quaketronics and shows up at his local Perl Mongers and Dorkbot meetings.

Photography by Sam Murphy (C. F. G) and Tom Anderson (D. E)

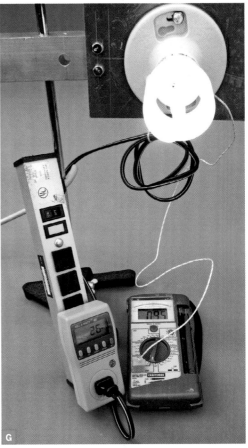

## TOO HOT IN THE CFL

When I suspected that overheating killed my CFL bulb, I cracked it open to look at the electronics. I built a tool to open the base by drilling a 1" hole in a block of wood. This holds the bulb by the base. Using a screwdriver as a lever, I pried apart the seam in the plastic base. To make it easier, I created a fulcrum from a scrap of aluminum angle. With practice I was able to separate the base without much effort. The CFL base easily snaps back together, and the light still works (Figures D and E).

⚠ CAUTION: Wear gloves and safety glasses if you open CFL bulbs. Don't touch the spiral glass tube; pushing the glass won't persuade the plastic to move. Work in a well-ventilated area such as outdoors or in a garage.

The bulb contains mercury, and should be disposed of according to the directions at lamprecycle.org. Don't fear the mercury — if the bulb breaks, the 5mg of mercury in it is unlikely to hurt you because it's in metallic form and it's a tiny amount. Just take care not to touch it or breathe it, air out the room for 15 minutes or more, and follow cleanup instructions at epa.gov/mercury/spills/index.htm#fluorescent.

The circuit in the base of the CFL is called a ballast (Figure F). It regulates the current drawn by the bulb; even as the AC voltage fluctuates, the current stays the same. It uses a 20µF 200V electrolytic capacitor with a temperature rating of 105°C (221°F). The capacitor is expected to last 10,000 hours — but for every 10°C (18°F) of heat above its rating, its lifetime drops by a factor of 2.

You can measure the temperature inside your CFL bulb with a thermocouple. Sears sells a voltmeter that comes with a thermocouple and measures temperature (Figure G). Take apart the plastic CFL base, drill a small hole in it, and put the thermocouple inside. (Because the thermocouple is made of wire, take care that it doesn't short-circuit the electronics.)

After it warms up, you'll find that the temperature in the base of the CFL varies when you change the orientation of the bulb. Measure it in a variety of positions, and try it in a totally enclosed light fixture. I found that the enclosed fixture causes only a small rise in temperature, but there's a large increase when the bulb operates upside down.

# Heat Exchanger

**Fresh air without the energy loss.**

BY CHARLES PLATT

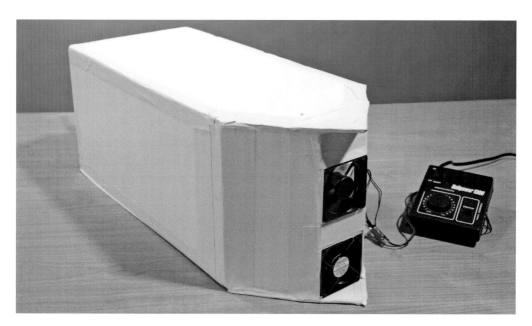

**A**n ideally energy-efficient house should be tightly sealed, to hold cool air inside during the summer and keep it out during the winter. The trouble is, we need to circulate some fresh air to remove odors, bring in oxygen, and reduce the risk of mildew and mold growth.

Is there a way to move air in and out of a house while minimizing the heat moving in and out?

One simple gadget can do it: a heat exchanger, aka "heat recovery ventilator." Instead of letting air run in and out freely, the exchanger uses two small fans to draw incoming and outgoing air through parallel interleaved ducts. The two flows don't mingle, but heat passes between them through the thin metal walls of the ducts.

In the winter, warm air going out through the heat exchanger gives up its heat to cold air coming in, and during the summer, cool air blown out through the heat exchanger steals heat from hot air coming in, so that by the time the incoming air enters the house, it isn't hot anymore.

Heat recovery ventilators are cheap to run, because they only contain a couple of fans. But they can be costly to buy, retailing for $450 and up.

Here's how you can build your own for radically less, $50 to $100, depending on how many materials you already own. This is my first design attempt, and it works, but I don't pretend that I optimized it. Feel free to make it better.

I am very grateful to MAKE intern Eric Chu for tackling the hard work of fabrication and testing, using plans that I drew (Figures A–G, pages 58–59).

## Design

These are the crucial design features to get the most out of a heat exchanger:

» Interior panels should have maximum surface area relative to volume.

» Panels should be made from thin, thermally conductive metal.

» Incoming air and outgoing air must move in opposite directions.

» Fan speed should be adjustable, to maximize heat transfer.

Since aluminum conducts heat very efficiently, I decided to make the panels from aluminum foil stuck to wooden frames, with holes drilled in the edges of the frames to allow air to pass through. Cheap computer fans are fine since they're quiet and don't use much power. Since this unit is merely providing some moderate ventilation rather than heating or actively cooling roomfuls of air, the flow rate can be low.

You might assume that slower airflow allows greater opportunity for heat transfer between air going out and air coming in. Theoretically, this should be true, but in practice other factors play a part, such as heat penetrating or escaping from the box containing the unit. When we tested our exchanger on the cold night air, we found that higher fan speeds actually warmed the incoming air more. Perhaps this is because faster-moving air lengthens the temperature gradient along the heat-transfer path and keeps the box containing the unit from getting cold.

Where, exactly, is the sweet spot? I suggest you build the unit and adjust the fan speed experimentally to find out.

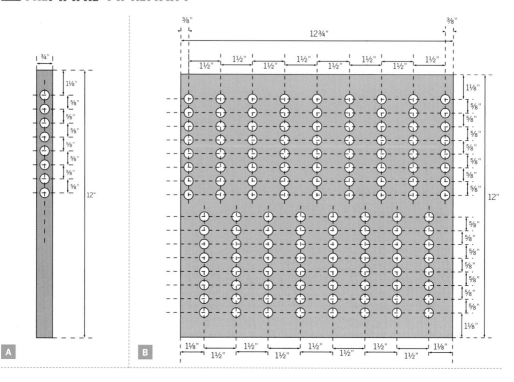

**A**

**B**

## 1. Cut and drill the hardwood.

Cut the ¾" square wood into 18 lengths of 12" and 18 lengths of 24". The shorter pieces should be exactly 12", but it's OK if the thickness of saw cuts causes your 24" lengths to be slightly short.

Starting 1⅛" from one end of each of the 12" sections of wood, drill 8 centered holes ⁷⁄₁₆" in diameter, spaced ⅝" apart (Figure A). Reduce the diameter if your wood is undersized.

## 2. Make the frames.

Use the hardwood pieces to make 9 rectangular frames, with the holes in the 12" pieces running in and out of the frame, rather than front to back, and positioned upside down relative to each other (Figure C). Use fast-setting epoxy with 4 picture-frame clamps to hold the corners. Number the frames to identify them.

## 3. Make the end panels.

Cut 2 pieces of masonite (or ABS, or plywood), each measuring 12"×12¾". Drill each piece with columns of ⁷⁄₁₆" holes following the plan in Figure B.

The holes in your frames must align exactly with the columns of holes in the end panels. To make sure that they do, clamp a frame to the panel in each column position, and then drill the holes into the panel through the holes in the frame, to make them match.

## 4. Apply foil.

Cover both sides of frames 2 through 8 with rectangles of foil, using double-sided tape. Cover frames 1 and 9 with foil on only one side, the inward-facing side.

**NOTE: Small air leaks around the edges of the foil are not very important, but tears in the foil will allow too much leakage.**

## 5. Install the panels.

Attach frame 1 along one side of an end panel, matching the holes and facing the foil inward (Figure D). Use three 1"–1¼" #6 flathead wood screws at top, middle, and bottom. Similarly join the other end panel onto frame 1, upside-down relative to the first end panel. Then attach all the remaining frames into place, orienting and aligning their holes as you did with frame 1 (Figure E). When you get to frame 9, its foil side should face inward (Figure F).

## 6. Enclose it.

The hard part is over. Now box up your frames by cutting top, bottom, and side panels, and taping them into place. For our test version we just used foam board, but for something sturdier to install in a house, screw or glue masonite panels, then clad them with foam board for insulation.

## 7. Mount the fans.

Make 2 manifolds for the fans, one sealed over the top set of holes (Figure G), and one over the bottom set, on the end of the box that will sit inside the house. One fan will blow inward, while the other blows outward (Figure H). We made a single trapezoidal hood by cutting, bending, and taping foam board, and then used another piece of foam as a divider inside.

Fit the fans snugly in the manifold ends as shown on page 56. We taped around them, for airtightness more than to hold them in place. That's it!

## Time for Testing!

First, measure the temperature outside and inside. For a good test, you need at least a 20°F difference between the two. Place the heat exchanger in an open door or window and block the open space around it to prevent stray air currents.

We aimed ours out a screened window masked with more foam board. The screen may have increased the amount of exhaust air that doubled back and was drawn inside again. To minimize this, you could divert the outgoing air off to the side with a small baffle or shroud.

Run the fans, and measure the temperature of the incoming air. It should gradually move closer to the inside temperature. If you have an adjustable fan-speed control, try different speeds to find the one that warms incoming air the most.

Our exchanger worked best when the fans ran at 100%. Here are our results at that speed:

| | TRIAL 1 | TRIAL 2 | TRIAL 3 |
|---|---|---|---|
| Outside temperature | 42.6°F | 45.3°F | 41.0°F |
| Inside temperature | 71.4°F | 67.1°F | 67.4°F |
| Incoming air temperature | 62.2°F | 58.2°F | 58.1°F |

Your heat exchanger will be especially useful in a home that's fairly airtight, with good draft seals around doors and windows. In the summer, it can save power by providing ventilation without losing the coolness from your recirculating air conditioner.

In the winter, it's ideal in conjunction with an unvented gas heater. This type of heater retains all its thermal energy inside the house, while the heat exchanger gets rid of carbon dioxide and brings in oxygen. Together they form a perfect partnership.

➕ Download the heat exchanger plans at makezine.com/18/heatexchanger.

Charles Platt builds prototypes of equipment for medical research and is a contributing editor to MAKE.

# Off-Grid Laundry Machine

## This think-small washer needs no electricity or running water.

BY MICHAEL PERDRIEL

**A** couple of years ago, I decided to concentrate my design research on devices that would be useful to poor families in developing countries — easy-to-make tools that address a specific need without disrupting the local economy, culture, or environment.

Here's one of my designs: a manual clothes washer that does a load of laundry in about 20 minutes using no power other than muscle. It's portable, so you can carry or wheel it to a water source, and if you wash with biodegradable soap, the wash water can easily go to a garden afterward.

They're now using the washer in Hyanja, Nepal, where I collaborated on designing a localized version (*see page 67*). It's also a neat design if you're living off the grid by choice in an industrialized area, or just conserving water and power.

Illustration by Alison Kendall; photography by Michael Perdriel

## MATERIALS

**10gal plastic bucket** or any deep waterproof container big enough to hold clothes and with a mouth at least 12" wide. Handles and a lid are also useful.

**1×3 hardwood boards,** 40" (1) and 20" (2) A 1×3 measures ¾"×2½". I ripped a 1×6 in two.

**¼" plywood,** 16"×16" to strengthen the frame's corners. You can also use strong plastic like plexiglass, or any rustproof sheet metal you can cut.

**1" hardwood dowels,** 3' long (4)

**Wooden tool handle** for rake or hoe

**¼" braided poly rope,** 50' or other fairly stiff rope

**⅛" braided nylon parachute cord,** 20' or other thin, strong, waterproof rope that holds a knot

**Flexible netting,** 12"×30", with 2" holes or smaller I reused a discarded tennis net.

**PEX plastic tubing,** ⅝" OD, 10' lengths (2) or other semi-rigid tubing; in Nepal we used bamboo.

**Package of 4" cable ties** or you can use thin rope

**¾" #6 stainless steel screws** (24 or so)

**2" deck screws** (4)

**Waterproof glue**

**1"-diameter copper pipe slip connectors** (2)

**Aluminum bars,** ⅛"×¾"×3" (2) or any other rust-resistant metal

**⁵⁄₁₆"×3" screw eye**

**¼" bolt,** 1¼" long, with washers (2) and nuts (2)

**Nut to fit the ¼" bolt** that also just fits inside the screw eye

## TOOLS

Screw gun or variable-speed drill
Screw bits
Drill bits: ⅛", ⁵⁄₁₆", ¼", and 1½" hole saw
PVC socket flange, ¾" hole I found one branded "Nibco" in the electrical supplies department.
Hacksaw, and handsaw or jigsaw
Pliers, and adjustable wrench or locking pliers
Rasp
Flathead screwdriver
Utility knife
Hammer and a mallet, or hardwood block
Tape measure
Sandpaper
Vise
PEX tubing cutter (optional) saves time
Cable tie fastener (optional) saves time

**CLEAN ENERGY:** Laundry machine prototype in Nepal (top), and a small, cartable version of the machine built as a woodworking project (bottom).

## » Inciting Agitation

The washer consists of 3 main components: a container, a net bag, and a lever-driven shaft mechanism held in place by a simple wooden frame.

The key component is the net bag, which is designed to hold, squeeze, and agitate the clothes. The middle of the net bag is a wide, open cylinder of flexible mesh netting. End-capping the cylinder above and below are semi-rigid cones made from short plastic pipes strung together with rope.

Both cones point upward, so the bottom cone sticks up through the clothes and prevents them from balling together.

While the washer is in operation, the top cone holds fast while the bottom cone is pulled up and down by the shaft, carrying the clothes with it.

Each pump of the lever handle pulls the clothes up out of the water, squeezes them out between the nested cones, and releases them back down. The lever's 40" length provides mechanical advantage for easy operation.

These instructions show how to build a bare-bones device for less than $50 using materials from any home supply store. You can modify the design to suit available materials and your skill level. A machine of this size can handle only small loads up to 5lbs, but the ones we made in Nepal were larger, and I think that one could be made 2 or 3 times larger and would still be easy to operate. I also built a fancier, wooden version that's towable, with wheels and a barrel-style container.

## 1. Make the wooden frame.

**1a.** Saw the 1×3 lumber into one 40" length and two 20" lengths.

**1b.** About 1" from one end of the 40" board, drill 2 overlapping 1½" holes and rasp the edges to create a 2"×1½" oval. Bevel and sand the hole on both sides of the board. This is your vertical support.

**1c.** In one of the 20" lengths, drill a centered 1½" hole 8" from one end; this is the top horizontal plate. To help tie it to the wash bucket (optional), cut notches as shown in Figure A: 1" from the end nearest the hole, use a rasp or drill bit to make 2 notches ½" wide, one in each edge of the board. Make 2 more notches 4" from the opposite end of the board, so that the hole is centered between the pairs of notches. The plain 20" board is the base plate.

**1d.** Saw the plywood into four 8"×8" pieces. Cut each piece into a ¼ circle shape (or a right isosceles triangle) with two 8" sides at right angles. Sand the edges smooth. These are your gussets (Figure A).

**1e.** Butt the 20" base plate at a right angle against the bottom of the vertical support (the end without the hole). Drill two ⅛" pilot holes, then glue the 2 pieces together and fasten with deck screws (Figure B).

**1f.** Butt the top horizontal plate at a right angle against the vertical support, facing opposite the base plate, 17" from the bottom and with the notched end away from the support. Drill 2 more ⅛" pilot holes, then glue and fasten with deck screws (Figure C).

**1g.** With ¾" screws, screw and glue the gussets at each right angle, one on each side, to reinforce the joints (Figure D). The top gussets should stand up, not hang down.

## 2. Make the drive shaft and handle.

**2a.** Secure the 1" dowel in a vise. Draw 2 lines on the end, ½" apart and equidistant from the center point. With a handsaw, cut notches along these lines 1¾" deep into the dowel, leaving a gap between the

2 notches of at least ⅜". Use a regular handsaw rather than a thinner hacksaw, because you want notches about ⅛" wide, to accommodate the aluminum pieces (Figure E, at right).

**2b.** About 3" up from the dowel's opposite end, drill a ¼" hole perpendicular through the center of the dowel, at any angle relative to the notches. This will be used to attach the bottom cone to the shaft.

**2c.** Now take the tool handle, which will be the handle of the machine. Drill a ¼" hole vertically into the bottom (tool end), as deep as the screw eye shaft is long.

**2d.** Fit one of the copper connectors over the drilled end of the handle; you may need to sand it a bit. The copper should fit easily but snugly onto the wood. It will act as a ferrule, to keep the screw eye from splitting the wood.

**2e.** To make the aluminum brackets, cut the ⅛"×¾" aluminum bar into two 3" lengths. Drill a ⁵⁄₁₆" hole through both pieces, centered ½" from one end. Drill them stacked together, so that the holes line up precisely.

**2f.** Put the other copper connector over the end of the dowel with the notches. Insert the aluminum brackets into the notches, hole ends out. Use a mallet or block of hardwood to carefully and evenly tap the brackets into the notches until only about 1½" of bracket sticks out from the dowel. This should expand the wood against the copper ferrule enough to make it snug (Figure F, at left). If not, you can secure the brackets more by covering the joint with epoxy or by drilling a hole through the ferrule and brackets and securing them with a screw or small bolt.

**2g.** Screw the screw eye into the hole in the end of the tool handle until the center of the eye is about 1" from the wood (Figure F, at right).

**2h.** To attach the handle to the shaft, fit the nut in the middle of the screw eye to act as a bushing, then put the eye between the 2 aluminum brackets and attach them with the ¼" bolt, using washers on the outside surfaces of both brackets sandwiched between nuts at each end. Tighten the end nuts to keep the joint together, but not so much that they impede movement (Figure G).

## 3. Assemble the cones.

**3a.** Cut the PEX tubing into 32 pieces 6" long and 32 pieces 1½" long. Each cone, top and bottom, uses 16 of each (Figure H). I used a PEX cutter, but a utility knife will also work.

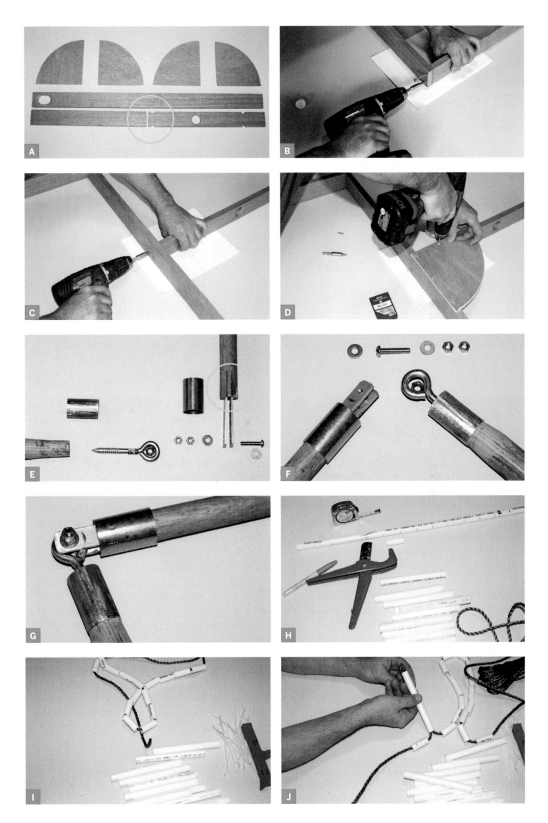

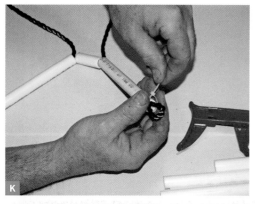

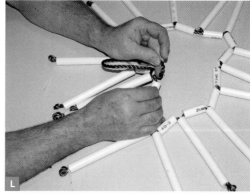

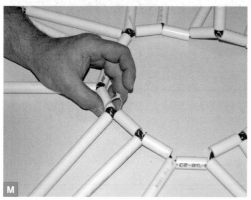

**3b.** Thread 16 of the short pieces of PEX tubing onto the ¼" braided poly rope, then pull the rope through until it runs about 18" beyond the chain of tubes (Figure I, previous page).

**3c.** Pull a loop of rope from between the first two threaded pieces of PEX and shove it through one of the 6" tubes. Feed this extra loop from the long end of the rope, maintaining the 18" tail at the other end (Figure J).

**3d.** Use a cable tie or thin rope to bind off the loop, leaving about ¼"–½" of the loop above the tube (Figure K). Repeat Steps 3c and 3d with 15 more 6" tubes, placing a long tube between the short tubes.

**3e.** Cut the rope, leaving about an 18"–24" tail, and thread the ends of the rope back around through all the short tubes again to close the circle. Tie the ends together in a tight square knot, and trim them (Figure L).

**3f.** Cable-tie or bind every other joint around the ring (Figure M).

**3g.** Thread a length of parachute cord or thin strong rope through all the loops and draw them together to form a cone (Figure N).

**3h.** Repeat Steps 3b–3g to make a second cone.

## 4. Attach the cones to the drive shaft.

**4a.** On the first cone, tighten the cord running through the loops and tie it so that the top hole is just large enough for the drive shaft to slide through smoothly. Hang the cone just under the wood frame's top horizontal plate by tying short lengths of cord to each side of the 1½" hole. The cone (actually, more of a frustum) should hang centered under the hole (Figure O).

**4b.** Use parachute cord to lash one long edge of the 12"×30" netting to the bottom of the cone. When you've completed the circle, tie the ends of the cord securely and use cable ties to reinforce the connection (Figure P).

**4c.** Slide the handle through the oval hole in the vertical plate and insert the shaft down through the hole in the top plate and the top of the cone, so that it hangs inside the netting (Figure Q).

**4d.** Take the second cone, and, using the cord threaded through the loops tie it around the shaft. Thread the ends of cord through the ¼" hole in the shaft (Figure R).

**4e.** Weave the ends of the cord back around through

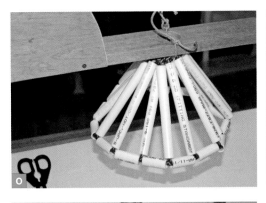

O

P

Q

R

T

S

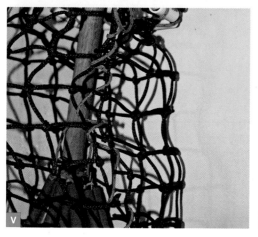

the loops, to help keep them arrayed evenly around the shaft. Finish by tying securely (Figure S, previous page).

**4f.** Lash the bottom of the net to the bottom cone and tie the ends securely. Reinforce the connection with cable ties (Figure T).

**4g.** Drill four ⅛" holes 90° apart through the sleeve of the ¾" PVC socket flange. Push the flange over the bottom of the shaft and into the bottom cone. Secure the flange to the dowel with #6 stainless steel screws through the ⅛" holes, then cable-tie the flange's 4 large holes to 4 of the long PEX tubes in the cone, spaced around evenly (Figure U). This will stabilize the cone and keep it from turning sideways during agitation.

That's it — you're done!

## Wash Time!

To use the washer, load clothes into the opening on the side of the net. Where possible, hook them through the top row of holes in the net, to keep them all from bunching at the bottom.

To close the bag, tie a 24"–30" cord to the bottom of the net near the opening, and lash the opening shut, tying with a loop knot at the top (Figure V).

The machine is easiest to operate from a seated position, with one foot on the base to stabilize it. Lower the bag and cones into the container filled with water and soap or detergent (Figure W), then pump the handle to agitate the clothes until they're clean (Figure X), usually about 15 minutes.

You may want to experiment with using both hands for greater action. Switch to a bucket of rinse water and pump until clothes are free of soap.

Finally, untie and "unzip" the bag cord, remove the clothes, and hang them to dry.

Michael Perdriel is a Pittsburgh-based maker, sculptor, furniture maker, and product designer with an interest in off-the-grid design. He has an MFA in industrial design from the University of Notre Dame.

FOUR-DOLLAR WASHING MACHINE: Nepalese villagers test a prototype of the off-grid laundry machine, manufactured locally for $4 using wood, jute rope, and bamboo tubes.

# OFF THE GRID IN NEPAL

I'm interested in designing products that benefit people who live "off the grid" either by choice or by necessity. Most of the world's laundry is done by young women and girls washing by hand. It's an effective way to get clothes clean, but it can be bad for both the women who do it and for the environment. Some women wash for a 6- to 7-hour stretch each week, which is hard on the back and joints, abrades the skin, and exposes them for prolonged periods to winter cold, waterborne parasites and pathogens, and possibly predators. Environmentally, the common practice of washing with uncontained water from a river, communal pool, or other source dirties a lot of water and releases it back to its source unfiltered.

I designed this machine to be built on-site, with locally available materials. I field-tested the design in Nepal in summer 2007, working with local craftsmen in Hyanja, a remote village outside of Pokhara. A local carpenter made the frame out of available wood, and he used precise mortise-and-tenon joinery instead of simple screws to construct the frame. We made the basket and cones from jute rope and bamboo, and the container was an inexpensive plastic barrel.

The Nepal washer was a bit larger than the one described in this article, and it satisfactorily cleaned a large load of clothes (including jeans) in about 15 minutes, which is probably half the time it would have taken without the machine. The total cost of producing our prototype machine in Nepal, for both materials and labor, was about $4.

# Making the Energy Garden

**An 18-month experiment in self-sufficiency.**

BY JULIAN DARLEY & CELINE RICH-DARLEY

FOOD+FUEL: The front yard of the Energy Garden at the height of summer 2007.

On Valentine's Day 2007, we immigrated from Vancouver, B.C., to Sebastopol in Sonoma County, Calif. To our surprise, the house we'd rented, sight unseen from an ad on Craigslist, had a double-sized lot. The third of an acre was in a very run-down state; it had been used as a Rottweiler and pig run by the previous tenants. Our request to turn the yard into a garden was granted by the landlord, and the Energy Garden was born.

For some years we had been concerned about climate change and energy depletion and how a liquid fuel shortage would affect the world's food supply system. With our colleagues at the University of British Columbia and the University of Kentucky, we were looking at small-scale farms to see whether they could be energy self-reliant.

The first thing that sprang to mind when we saw the yard was a question: can a family, in the middle of a town, be food and energy self-reliant, and even have some left over to share? Our garden would be an experimental demonstration. Its goal was to produce as much food and fuel as possible with as few outside inputs, especially petroleum, as possible.

We wanted to produce the "five Fs" in our garden: food, fuel, fiber, fertilizer, and feedstock — all things that we currently rely on petroleum for. We were also interested in the calorie. How many calories could we grow, both to eat and to turn into other forms of energy, and how could we demonstrate different methodologies of gardening to the public?

## Methodologies

We used John Jeavons' book *How to Grow More Vegetables* as a guide for the Grow Biointensive method, which advocates double digging of beds and helps calculate expected yields from areas of land. We also drew on permaculture methodology, including "guilds" of mutually beneficial plants, and crops with "stacking functions," where many parts of the plant are useful for different purposes.

We also staggered our plantings, following market

Photography by The Energy Garden Team

garden techniques, to keep the beds continually in use growing food, energy, or compost crops.

All of the garden work was done by hand. We started in March 2007 with forty 10'×4' beds, and a few weeks later we added three 30'×4' beds. Then we turned the whole front yard into ten 15'×4' beds. To use water most efficiently, all of this was drip-irrigated by a computer-controlled timer.

By the time summer came, the last open space of the backyard had been turned into a mandala garden with 3 concentric rings, using the layering method and planted with cover crops to wait for spring. By autumn, all the trees had guilds around them planted with berries, rhubarb, and shrubs.

A small section demonstrated "square-foot gardening," to show people that they could start small. And in summer 2008, the front side-yard demonstrated the "Do Nothing" method that some farms use to minimize soil erosion: we established a cover crop of clover and vetch, then cut it short and planted starts of quinoa into it.

In order to get plants started quickly, we built three cold frames and a temporary greenhouse out of hay bales and clear plastic.

## Growing Calories to Eat

Since we wanted to focus on calories and to show people the plants that they eat, we grew grains: wheat, amaranth, millet, quinoa, buckwheat, corn, and oats. We also planted traditional vegetables, such as tomatoes, broccoli, cauliflower, kale, cabbage, squash, watermelons, beets, and beans.

Over time we also learned what didn't work in the Energy Garden — such as potatoes, which were eaten by the gophers. There were wonderful old apple trees on the property, and we planted more fruit trees and berry bushes. Our medicinal and culinary herb beds did quite well. We grew echinacea, and our basil was the envy of the town.

## Growing the Soil

It may sound strange, but if you are to have a sustainable production garden, your first and last task is to grow your soil. This means building the organic matter as well as keeping up vital nutrients such as nitrogen. The traditional way to do this is by making compost and, if possible, keeping chickens. Making compost is not terribly difficult, but a good compost pile needs the right mixture of carbon-rich matter (for instance, straw) and other vegetable matter.

We found it very effective to keep the compost piles in the chicken run. This way, new kitchen scraps and other vegetables got thrown on top of the compost pile, and the chickens eagerly eat what they want while conveniently depositing their nitrogen-rich fertilizer into the pile. The resulting compost was like rocket fuel for the plants.

## Making Liquid Fuel

We grew oilseeds for biodiesel, high-sugar crops for ethanol, and cellulose-based crops for potentially making biofuels. We wanted people to see some of the crops that were being discussed in the news. For oil: Peredovik sunflowers, corn, canola, soybeans, and flax. For sugar: Dale sorghum and Jerusalem artichokes. For cellulose: switchgrass and miscanthus grass. It was great fun to watch these crops grow. The sorghum, for instance, touched the power wires at the front of the house — about 15 feet high.

We experimented with many energy machines, such as small wind turbines, solar panels, biogas digesters, and oil presses for biodiesel. Our most in-depth experiment was making ethanol both from excess apples and from sorghum juice.

After much searching we managed to find and transport a sorghum mill with which to squeeze the juice out of the fat, green sorghum canes. This machine weighs 400 pounds and is more than 100 years old (see Dale Dougherty's write-up in MAKE, Volume 17, "The Lost Knowledge Catalog").

With a 20-foot pole strapped onto the mill shaft, and about a dozen volunteers, we squeezed a couple of gallons of juice out of the sorghum canes on the lawns of O'Reilly Media. We boiled part of it down to make sorghum syrup, which is a bit like maple syrup. Another part we fermented to make ethanol. But thanks to lack of temperature control, and possibly pilot error, we produced nothing usable. Producing ethanol in quantity is a delicate job.

However, we did manage to properly ferment our apple juice and put it through the moonshine still that we'd bought. Voilà! About 1 liter of 150-proof alcohol. But before it could go into a gas tank, we would still need to filter it through zeolite (a volcanic rock), to remove the last water from the alcohol.

## Community U-Pick

Thanks to Northern California's temperate climate, the gardens could keep producing all year round. We built several machines to process all the food, including a solar oven and a solar drying cupboard. We were also given a grain mill.

But there was still no way that the two of us could process all the food that the garden was producing. So we started a community U-Pick for the 2008 harvest. About six families regularly came by and

Fig. A: The chicken run. We started out with 4 and grew the flock to 15 because the chickens were such good calorie producers. Fig. B: Cold frames with new starts. Fig. C: A modified apple press with a macerator that the Energy Garden team designed. Fig. D: The moonshine still distilling apple cider into ethanol fuel.

picked from the Energy Garden. They noted their harvest in a ledger book, which was tallied at the end of the month, and they paid farmer's market rates for the food.

## Conclusions

Trying to produce usable fuel is very difficult, and the Energy Garden really helped us all see why petroleum is so extraordinary and valuable, and why, given its dwindling future and climate effects, we should be trying to use as little of it as possible.

It was also a chance to make and use many tools, to see firsthand how to increase food security, and to understand fuel systems better.

Based on our experiments, we decided that at the home scale it's better to concentrate on growing food rather than fuel, while generating electricity with solar and wind power, and practicing energy conservation and efficiency.

In the end we would admit that, though the energy crops were crowd-pleasers and fun to grow, biofuel is, for now, better produced by using waste products — fallen, unused fruit for ethanol, waste cooking oil for biodiesel, and waste organic matter, including ruminant manure, for biogas.

In the year and a half that the Energy Garden was producing, our efforts grew in many directions beyond the garden. More than 1,500 people visited and were inspired. The knowledge, seeds, and tools we collected were passed on to the community group Daily Acts (dailyacts.org) in Petaluma, Calif., who are incorporating these ideas into an urban homestead project.

The Energy Garden was realized by many people, including paid staff and volunteers. There were always people wandering through, from the neighbor bringing snails from his garden for the chickens, to volunteers bringing their visiting relatives for a tour, to anyone walking by that our 4-year-old son could entice in for a visit. One thing that the garden did build — in addition to soil — was community.

Julian Darley is the author of *High Noon for Natural Gas*. He has master's degrees in science (social research and the environment) and journalism, and is currently researching sustainable decision-making. Celine Rich-Darley has a master's degree in design for the environment from Chelsea College of Art and Design. Celine and Julian co-founded Post Carbon Institute. They now live in the London area.

# Kitchen Garden Quantitative Analysis

I've been maintaining a backyard farm for five years, and it's grown each year along with my enthusiasm. This year I'm seeing if its 1,200 square feet (ft$^2$) of growing space can produce at least 50% of our household's food.

Doing this means calculating with a few key factors: the space you have, the nutrient and calorie intake you require, the crops you can grow, and the schedule you can maintain for planting, irrigation, and soil maintenance.

For someone like me, who has a "job job" and is also in graduate school, it's a lot to think about. But I've relied on two helpful books: David Duhon's nutrition and garden planner, *One Circle,* and John Jeavons' mini-farming bible, *How to Grow More Vegetables*.

There are three people in our household. Given our ages and activity levels, I require 1,800–2,000 calories daily, my husband needs 2,400–2,800, and my father-in-law needs 1,500–1,800. For other nutrients, I referred to the guidelines published by the Institute of Medicine's Food and Nutrition Board. With that, it's a matter of breaking out a spreadsheet, looking at possible crops, and working out what your garden's expected yields can supply.

In addition, you need to plan crop rotation. Some crops deplete the soil, while others, like legumes, replenish it with nitrogen and improve compost. To maintain soil health, you must rotate crops growing in each location between the two categories. Luckily, crop rotation can also work well with balancing nutrition.

For example, the many beautiful varieties of kale are all easy to grow and nutritionally dense. Just 1 cup (uncooked) provides more than your daily requirements for vitamins A, C, and K, and good amounts of other nutrients. Combine it each day with 1 cup (cooked) adzuki beans and 1 egg (from our chickens), and you have 48% of the protein you need, 70% of the dietary fiber, and at least half of most essential vitamins and minerals. The calorie total from these three ingredients comes to just 420, but does not include any oils, butter, onions, garlic, or herbs you cook with.

We all like kale in our household, so I figure we can eat it 3 times a week. This translates to 9 cups, or one whole plant (they're large). Each plant needs a 15" radius and takes 2 months to mature, so we can grow all our kale in 3 rotating beds measuring 60"×15", containing 4 plants each. Each month one bed is planted, one is growing to maturity, and one is harvested.

There are two ways to harvest the kale. You can dig it up all at once, then wash, chop, and freeze some, or you can harvest the plants continuously for about 1 month, picking just the lower leaves while newer ones grow in. Either way, when the bed is empty again, you dig up the weeds, amend it with compost, and replant with the next crop. Transplanted legumes are a good option, and the location you transplanted them from can become a new kale bed.

To provide us all with 1 cup of beans per day, we need to grow 900ft$^2$ for 3 months. But once the beans have dried, we can store them and free the space for a more diverse range of vegetables. The beauty of both kales and dry beans is that they grow in cooler months, which lets you dedicate the summer weather to more finicky specialty crops like squashes, tomatoes, and basil.

For irrigation, we use drip lines on a timer (see page 72), which is efficient and inexpensive, and put mulch on the soil to preserve moisture and suppress weed growth. Our next steps for water conservation would be collecting rainwater and using household graywater.

Each of these considerations, from nutrition to irrigation, is a science in itself. For our household, it has been meaningful to figure it all out, and to reap the rewards of delicious food that we have a deep connection to and understanding of.

➕ Download Pallana's sample Crop Planner spreadsheet at makezine.com/18/gardenanalysis.

Esperanza Pallana (pluckandfeather.com) is an urban farmer in Oakland, Calif., who consults with individuals and organizations looking to grow their own food.

# Drip Irrigation

**Grow healthy vegetables the automatic way, while using less water.**

BY ERIK KNUTZEN

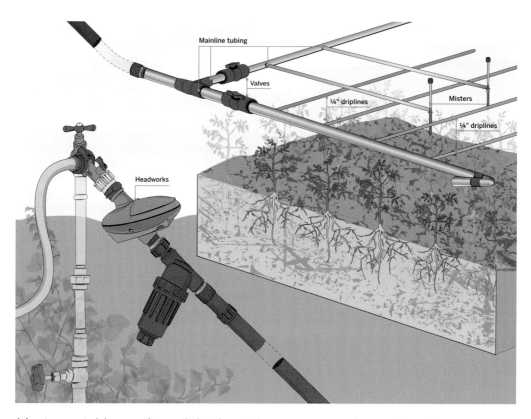

Mainline tubing

Valves

¼" driplines

Misters

¼" driplines

Headworks

Most vegetables prefer soil that's neither soggy nor dry, and earthworms and beneficial microorganisms do too. When there's too much water, these organisms drown. Too little, and you find yourself with dead plants and a reputation as a "brown thumb."

How much is too much, and how little is too little? How often should you supply water, and how can you remember to do so?

Drip irrigation answers these dilemmas, giving plants the perfect amount that they need to thrive, and saving water at the same time. Irrigation also keeps water off the leaves of the plants, preventing nasty maladies like leaf mildew, and you'll suffer fewer weeds by delivering water only to the plants that you want to grow.

The chief drawbacks to drip irrigation are cost, the use of plastics, and the time and trouble for

installation and maintenance. For these reasons, I believe that drip irrigation is best reserved for your vegetable garden. For the rest of your yard, try to find plants that are adapted to your climate and don't need supplemental watering.

In this article I'll explain how to assemble a typical layout to water a vegetable garden in a raised bed of quality soil.

The array of tiny plastic drip irrigation parts and supplies can seem confusing at first, but the principle is simple: you're simply piecing together a stretch of hose that leaks.

Illustrations by Alison Kendall

## Hose Threads vs. Pipe Threads

A word about the two types of threads you'll encounter with drip irrigation parts. Hose threads, used on garden hose and outdoor faucets, are more widely spaced than pipe threads. Drip irrigation parts may have either hose threading or pipe threading, the latter often being identified as "NPT," referring to the National Pipe Thread standard. If you try to attach a pipe thread to a hose thread, you'll strip the threads.

All parts recommended for this project have hose threads.

## Teflon Pipe Tape

To prevent connections from leaking, prepare them with teflon pipe tape. Before joining two threaded parts, wrap the tape around the male thread in the same direction that the second part will turn when you attach it. Hand-tighten all plastic connections; a wrench can damage the delicate threads.

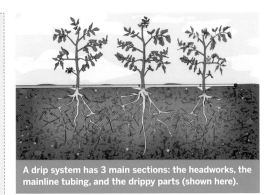

A drip system has 3 main sections: the headworks, the mainline tubing, and the drippy parts (shown here).

## MATERIALS
**For a 4'×8' raised vegetable bed. All the parts recommended for this project have hose threads.**

Backflow preventer
Teflon pipe tape
Hose Y connector
Battery-operated timer **Simple programmable ones cost about $40.**
Y filter with mesh
Pressure regulator, 25psi or below
Female hose beginning
½" mainline tubing **comes in rolls**
¼" mainline tubing **comes in rolls**
¼" dripline (32') with emitters every 6"
½" elbow fittings (3–4) **depending on terrain**
½" coupler fittings (1–2)
½" valves (2)
Figure-8 end fitting
¼" goof plugs (4) **but have extras on hand**
¼" hose barbs (4)
360° misters (2)
Stakes with riser (2)
Hold-downs (8) **and more would be better. You can also make your own from strong wire.**

## TOOLS
Hole punch
Thermos
Scissors or knife

## » The 3 Sections of a Drip System

A simple drip irrigation system consists of 3 sections: the parts near the faucet, which I'll call the "headworks"; the mainline tubing; and the parts that drip, which I'll call the "drippy parts."

The headworks perform 3 tasks: reducing water pressure, filtering your water supply, and turning the system on and off. The mainline tubing distributes the water to your vegetable bed, and the drippy parts drip, creating localized volumes of moisture beneath the soil. For your raised vegetable bed, you'll lay out parallel lines of emitter tubing to create a grid of evenly moist spots across the bed, and you'll add some misters for starting seeds.

## 1. The Headworks

Most drip irrigation suppliers offer kits that contain all the headworks parts that you'll need in one box, plus a few extras. If you plan your project carefully you can skip the kit, omit the extras, and save a little money by buying just the parts you need.

The headworks begin with the hose bib (aka "faucet") that you probably have in your yard. It needs to be in good condition, since you'll be leaving the valve open all the time. If it leaks, fix or replace it.

Starting at the hose bib, thread together the parts in the following order, remembering to add a few wraps of teflon tape around each thread (see "Drip Irrigation 101" above) before you connect it.

**1a. Backflow preventer (aka anti-siphon device).**
Required by code in many places, this part (item A, following page) prevents water from being siphoned from your irrigation lines back into your water supply. This keeps potential pathogens that might be lurking in fertilizer and soil from contaminating your drinking water. Many new homes may already have them installed on outside faucets.

A–F: Headworks include a backflow preventer, Y valve, timer, filter, and pressure regulator. G–K: Mainline with valve, couplers, and end fitting. L–S: Drip components include the dripline, hold-downs, and misters.

Backflow preventers come in plastic and brass versions; brass costs more, but lasts longer. They have an arrow etched into them, which must point in the direction of water flow.

**1b. Hose Y connector with valves** Y connectors (B) allow you to keep an additional garden hose connected for hand watering, dog washing, and other purposes. A brass one will last longer than a plastic version and is less likely to break or leak. The Y should be fitted with 2 small valves so that you can shut off either the drip system or the hose independently.

**1c. Timer** Nothing kills plants faster than inconsistent watering, especially in hot weather. A timer (C) takes the flakiness factor out of your gardening endeavors, getting water to those plants when you're still in bed shaking off a hangover. For this basic system you'll use a battery-operated timer, which you should be able to pick up for around $40. Get

one with an LCD display that lets you set custom watering times during the day.

**1d. Filter** The small apertures of drip emitters are easily clogged by dirt, rust, and hard water deposits in your water supply. Drip irrigation filters (D) have a mesh inside that catches this gunk. For the type of emitters you'll be using you need to get a Y-type filter, so named because it has 3 threaded fittings: 1 input that you'll connect to the output of the timer, 1 output that you'll connect to the pressure regulator, and 1 valve or cap at the bottom of the filter that you'll open periodically to flush out sediment that may have accumulated in the mesh.

**1e. Pressure regulator** Drip irrigation systems operate at low water pressure, usually less than 25psi. Most municipal supplies deliver water at a higher pressure than this. Without the regulator (E) to step down the pressure, the connections in your drip system will leak or burst. Pressure regulators

Photography by Erik Knutzen

are specified by output pressure, measured in pounds per square inch (psi) and flow rate, measured in gallons per minute (gpm).

For your drip emitters, you need a regulator rated at 25psi or less. Most entry-level pressure regulators ($6 or so) can handle a flow rate of 0.5–0.7gpm, which is more than enough for a small vegetable bed. The pressure regulator should have hose threads, and will have a little arrow on it to show the direction of flow.

### 1f. Female hose beginning (FHB) and ½" mainline tubing

You're almost done assembling your headworks. The last step is to stick the ½" mainline tubing (G) into the female hose beginning (F). This is easier to do before you attach the FHB to the pressure regulator.

Mainline tubing, made of black, flexible polyethylene, is the main artery of your drip system, carrying the water from the headworks out to your garden beds. You can cut it with scissors or pruners. Mainline tubing is inserted into compression fittings, which means it's jammed in, not screwed in. The FHB has a compression fitting on one side and a hose thread on the other.

Once you've inserted the mainline tubing into the FHB, simply thread the end of the FHB onto the end of the pressure regulator, and you're done

assembling the headworks — but not quite ready to lean back and crack a beer just yet. You still have to place the tubing to distribute the water to the vegetable bed.

**TIPS: Compression fittings can be difficult to work with, especially on cold days, when the plastic is stiff. To make it much easier, put some boiling water in a thermos. Stick the tubing into the boiling water for a few seconds and then insert it into the compression fitting, rocking it back and forth to make sure it goes in all the way.**

**There are other kinds of proprietary fittings, such as Smart-Loc, that cost a bit more but are easier to assemble and disassemble than compression fittings. Consider this alternative if you think you may need to disassemble or reconfigure the system subsequently.**

## 2. Mainline Tubing and Fittings

**2a. Mainline tubing** Your mainline (G) should be protected from the sun as much as possible to extend its lifespan. (It's also ugly, and covering it will make your garden look more like Versailles). Burying it in mulch is an ideal way to prevent damage from dogs, people, and lawnmowers.

Roll out the tubing gently to prevent it from kinking. And if you're working on a hot day, remember that it can contract when it gets cold. Leave a little bit of slack to prevent problems later.

**2b. Elbow and tee fittings** Mainline tubing is pretty flexible, but if you need to add a branch or make a hard turn, you'll need an elbow or tee fitting (H). Like all the parts you'll be working with after the headworks, these are compression fittings.

**2c. Couplers** Most likely you'll thrust a shovel through your tubing at some point, or you'll trim it too short while you're laying it. Have a few couplers (I) on hand, to splice together broken lines.

**2d. Valves** Use 2 valves (J) to switch between the drip emitter tubing and the misters.

**2e. End fittings** Once you've placed the mainline tubing along the short end of your vegetable bed, turn on the water to flush out dirt that may have found its way into the fittings.

Next you'll cap off the far end of the mainline tubing by kinking it, using a figure-8 end fitting (K, T) to hold it. Insert the end of the mainline tubing into one hole of the figure 8, bend it back a few inches, and insert the end again into the other hole. Turn on the water again and check for leaks.

## 3. The Drippy Parts

With the mainline tubing run out to the vegetable bed, you're ready to insert the ¼" dripline that will

actually water the bed. Start by running the dripline straight down the long side of the bed at even intervals spaced at about 9"–12" on center. If your bed is 4' wide, 3 or 4 driplines should be enough.

**3a. ¼" dripline** For intensively planted vegetable beds less than 15' long, use ¼" dripline (L) with emitters (drippy points) embedded every 6".

**NOTE: Dripline is not the same as "soaker hose," which is made out of recycled tires and leaks water along its entire length. While soaker hose is cheaper and easier to find, it tends to distribute water unevenly along its length.**

To install dripline, simply cut it to length, insert a "goof plug" (M) at the far end to plug it, and insert a double-ended hose barb (N) to connect the other end with the mainline tubing. Use your hole punch (O, U) to make a small hole in the mainline tubing, then insert the hose barb into the hole.

**3b. Hold-downs** To keep each dripline in place, use 2 or more hold-downs (P, V). You can cut 2 lengths of wire, each 24" long. Bend them in half to create an inverted U shape and use the pair to hold the dripline flush with the soil. I like to use chain-link fence bottom wire, but coat hanger wire will also work. You can also purchase hold-downs.

**3c. Misters** Drip irrigation is great for established plants, but its localized moisture won't work for starting seeds. You could hand-water until the seedlings are established, but I use 360° misters (Q, W) that come on a little stake (R) with a 9" riser.

To connect the mister, use ¼" mainline tubing (S) with a barb at the end. Insert the barb into the ½" mainline tubing as you did for the dripline.

While you can run misters and dripline at the same time, you'll probably want to alternate, using the misters to get the seeds going and switching over to the dripline once the plants are established. To do this, connect the misters to a separate section of ½" mainline tubing controlled with a valve, as shown in the diagram on page 72.

## Planting

Once you've installed your drip system, turn it on before you plant. Note the irrigation pattern at the surface and, after running the water for at least ½ hour, dig down and take a look at the underground moisture pattern.

If you're transplanting seedlings, you'll need to make sure to place the roots so that they pick up water from the dripline (X). Plants with larger roots, such as tomatoes, can be placed farther from driplines; carrots and beets need to be placed closer. But all plants need moisture, and especially when dealing with small seedlings, beware of dry spots at the surface of the soil.

## Mulch Mulch Mulch

Once your seedlings are a few inches tall, apply organic mulch in the form of leaves, finished compost, straw, grass clippings, or wood chips (Y, Z). Mulch conserves water, makes your plants healthier, and protects your drip tubing from UV sunlight damage. Mulch ain't optional — if you want healthy plants you need to mulch!

## Watering: When and How Long?

Many variables determine when and how long to water: temperature, humidity, root depth, and hours of sunlight, to list a few. For most mature vegetables, you need to moisten the soil to a depth of 2'. Younger vegetables require less.

One objective way to determine whether you're watering long enough is to run your system and simply dig a hole and see how deeply the water penetrated. Despite a lot of advice to the contrary, recent research indicates that frequent light watering is better than infrequent deep watering, but this is a highly divisive topic among gardeners. Depending on the time of year, the heat, and the humidity, you'll probably need to run your drip system 10–40 minutes each day. Plants prefer to be watered in the early morning hours.

## Maintenance

Check your system while it's running, at least once a week. Occasionally an emitter or other connection will pop out under pressure, or be kicked out by kids or dogs. You'll also need to replace the battery in the timer periodically.

Take time to unscrew the cap on the bottom of the filter, and run the water to flush it out.

Having an automatic system is not a license to ignore the garden. Bad things happen — pests and bugs, in particular — when you don't pay attention to your veggies. Go out and visit them. To prevent your plants from getting waterlogged, turn off the system during rainy spells.

If you live where the ground freezes in winter, you need to protect your drip system from ice damage. You need to bring in the headworks, any valves, misters, and emitters. Mainline tubing can be left in place, but should be drained or blown out with an air compressor. Then cap the beginning of the mainline tubing, or tie a plastic bag around the opening. Fittings are especially vulnerable to bursting, so make sure you lift each one to drain out the water.

## Critters

Gophers and some breeds of dogs are notorious for treating driplines as water-filled chew toys. In a raised bed you can prevent this by lining the bottom with hardware cloth, which is actually a wire mesh.

## Going Bigger

For larger plantings, get what farmers use: T-Tape or similar drip tape. Your headworks assembly will be the same, with the exception of a lower-pressure regulator. You'll still use ½" mainline tubing, but the fittings for T-Tape are slightly different. T-Tape is a specialized product you won't find at a big box store, so I recommend getting a kit from an online supplier that will have all the parts you need.

While less durable, drip tape is more economical than ½" dripline. For small vegetable plots the ¼" dripline makes more sense, since, unlike drip tape, you don't need to get it in large quantities.

Erik Knutzen is the co-author, with Kelly Coyne, of *The Urban Homestead*, a guide to growing food, fermentation, graywater, and more. Erik blogs at homegrownevolution.com, a green living resource that provides urban self-reliance tips and tricks.

# Vermicomposting: Make Your Own Worm Bin

**Let hungry, squirmy wigglers take out the household trash.**

BY CELINE RICH-DARLEY

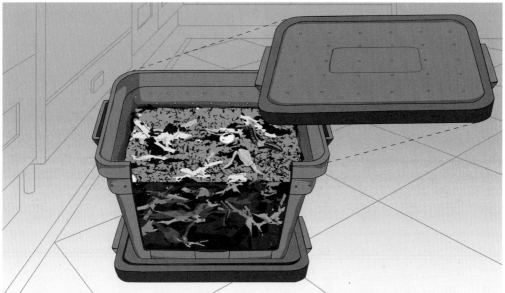

For three years I lived in a house with an outdoor composter. Then my family moved to an apartment in San Francisco where there was neither composter nor green box pickup! What was I going to do with all my kitchen scraps? I didn't have the heart to put them in the landfill garbage or down the garb-o-rator.

According to the EPA, in 2007 organic-based materials continued to be the largest component of municipal garbage in America: 33% was paper and cardboard, and 25% was yard trimmings and food scraps. I could put my paper and cardboard in the recycling bin, but without a yard, how could I recycle my apple cores, cabbage trimmings, and eggshells? Then I remembered worms.

Worm composters are great for apartments. No matter the climate or the size of your home, vermicomposting is good for you. Well, good for your plants. If you have children, there's the added advantage that most kids love worms (it's genetic), despite the fact that they're not very cuddly or furry.

My 4-year-old son is fascinated by worms: from our outdoor composter he'd already learned that worms turn kitchen scraps into soil as if by magic.

Soil is extraordinary stuff, and despite the fact that it's as vital as water, it's still not fully understood by scientists. But we do know that we're losing soil to erosion and runoff, and that composting can help restore soil, save landfill space, and reduce greenhouse gas emissions.

Worm composters are simple to build and easy to manage. Your worms can convert 5–6 pounds of food scraps a week into 10–15 gallons of compost a year.

Worm compost and worm tea (the drippings that collect in a tray at the bottom of the composter) can be used to fertilize both indoor and garden plants. Worm compost is higher in nutrient value than regular garden compost.

I'm looking forward to happy plants and less waste in my garbage cans, and so can you!

Illustration by Alison Kendall; photography by Julian Darley

## MATERIALS

Dark plastic storage box, 10gal–20gal in volume
Lids for the box (2)
Newspapers
1qt soil from the garden not potting mix
1gal water
Food scraps
Red wiggler worms (*Eisenia fetida*) aka redworms,
    brandling worms, or tiger worms. Buy them from a
    garden supplier or worm farm. Try searching online
    for "vermiculture supplies" or "vermicomposting."
Brightly colored tape
Popsicle sticks (2)

## TOOLS

Drill and bits: ¼", ¹⁄₁₆"
Trowel
Scissors

### » 1. Make drainage holes.
In the bottom of the plastic box, drill about
20–25 evenly spaced ¼" holes (Figure A).

### 2. Provide ventilation.
Near the top of the box, drill 2 rows of ¹⁄₁₆" holes
(Figure B). In the lid, drill 30 or so evenly spaced
¹⁄₁₆" holes (Figure C).

### 3. Prepare the bedding.
Shred newspaper or office paper. Use a paper
shredder if you can, or cut the newspaper into
roughly 1" strips with scissors.

Moisten the shredded paper with water and let
it soak in for a few minutes. This can be done in
a separate bucket to reduce the mess. The paper
should be damp but not soggy; squeeze out any
excess water and then put it in the plastic box.
Fluff it up so that there is air between the paper,
and fill the plastic box about ⅔ full (Figure D).

Add the soil to the plastic box (Figure E, next page).

### 4. Add worms.
With the trowel (or your hand), dig halfway through
the bedding in the middle of the box. Tuck your red
wigglers into their nice, moist bed (Figure F) and
then cover them with the bedding (Figure G).

### 5. Feed your worms.
Use tape and popsicle sticks to make an X-shaped
marker, or find some other object to use as a marker.
You'll put it on the bedding over the scraps, so you
can identify where you placed the food most recently.

Dig a hole in the bedding material in a corner of
the plastic box. Place a small amount of kitchen
scraps in the hole, cover it with bedding, and then
place the marker on top (Figure H). Worms like any
fruit, vegetable, or grain/bread. They also like coffee
grounds and filters, tea bags, crushed eggshells —

stuff you usually put down the garbage disposal.

Treat your worms like they're vegan — don't give them meat, fish, or dairy. They can eat these foods, but the bin will get smelly and attract pests. Also avoid oils, salt, and animal poop, and go easy on the citrus as it contains limonene, a compound toxic to worms.

It takes a while for the worms to get going, so don't be too impatient. As they multiply, they'll consume your kitchen scraps faster.

After a few days, check the bin. If the kitchen scraps are mostly gone, put another batch of scraps in. Put them to the right of the first batch, and then move your marker over to cover the new spot. Continue like this clockwise around your new compost box.

## 6. Situate your composter.

Your new worm composter can live in many places in your home: under the sink, in the laundry or storage room, even on the balcony. Your chosen spot should have good ventilation, easy access to collect the compost tea, and a suitable temperature. The best temperature for worms is between 55°–77°F year round, so make sure they won't freeze or fry.

Place the second lid under your new worm bin to collect the drips that will become compost tea.

## 7. Monitor your composter.

**» Moisture:** If the contents seem too dry, add a little water. If too wet, add shredded newspaper.

**» Smell:** The worm composter can become anaerobic (deprived of oxygen) if there is more food than the worms can eat quickly. If that happens, don't add scraps for a week or so. Give the worms a chance to catch up.

Also add more bedding (damp, shredded papers). Make sure there are enough ventilation holes in the container, and fresh air around the container. Fluff the bedding. If you leave it alone awhile, the situation should correct itself.

**» Fruit flies:** Make sure that food is buried and covered with bedding to avoid fruit flies arriving.

**» Worms dying or escaping:** Check the moisture content of the bin: if it's too wet, add bedding; if too dry, add water. If the contents look brown all over, then it's time to harvest your new soil.

**» Tea tray:** If the tray has a lot of brown sludge in it, scoop it into your watering can. Fill the can with water and let it steep for a day, stirring occasionally. Then water your plants with this highly nutritious compost tea fertilizer.

**» Harvest time:** When all the bedding is gone and your composter smells like a fresh forest (usually after 3–5 months) it's time to harvest.

It's better to harvest too early than too late, which can kill your worms. Any bits of food left over can be put back into the next worm composter iteration.

## 8. Harvest your soil.

**» Quick and messy method:** To separate the worms from the compost, empty the contents of the worm composter onto a tarp or old plastic tablecloth. Worms hate light and will wiggle into the pile. Wait a few minutes.

Then with your trowel or your hands, remove the top layer of the compost pile until you see worms. Then wait again, be patient, and continue removing the compost. Repeat until there are lots of worms in a small pile. All the worms can go into the next iteration of the compost box, or half can start another compost box.

**» Slow and neat method:** Make a second, identical compost bin by repeating Steps 1, 2, and 3. Take the lid off your first, full composter, and place the second bin directly on the compost surface of the first. Then repeat Step 5, putting kitchen scraps in the second box, and put the lid on the second box. In 1–2 months, most of the worms will have moved upstairs to find the food there. The first (bottom) compost box will contain mostly vermicompost.

**NOTE: Red wigglers are not native to North America. They are an invasive species in many areas, so don't dump worm-containing compost in natural areas; this could end up displacing the native worms.**

➕ More information about vermiculture can be found at ciwmb.ca.gov/organics/worms.

Celine Rich-Darley loves gardening. She has a master's degree in design for the environment and is the co-founder of the Post Carbon Institute. She lives in the London area.

## WORM FACTS

» The best worms for vermiculture (worm composting) are *Eisenia fetida* (striped) or their cousins *Eisenia andrei* (not so striped), aka redworms, red wigglers, tiger worms, or manure worms.

» Worms need grit for their gizzards to grind up and digest food, because they have no teeth.

» Worms need a moist environment because they breathe through their skin, which must be moist in order to breathe.

» Worms can eat about half their weight every day. Therefore, if you produce ½lb of kitchen scraps a day you'll need 1lb of worms. There are about 500 worms in a pound.

» Worms hate the light.

» Ideally worms like the temperature to be between 55°–77°F (13°–25°C). However, they can tolerate 40°–80° F. If they get too hot or cold, their activity slows down. Try not to kill your worms! Protect them from overheating (above 85°F in their box) or freezing.

» Worms need oxygen to live and they produce carbon dioxide. Your composter needs to be in a well-ventilated area.

» Worm castings (worm poop) are toxic to your worms. That's why it's important to regularly harvest your compost.

» Worms are great barometers. Before and during any low-pressure system such as a thunderstorm, worms like to crawl up and around the lid of the worm composter.

## THE REDWORM LIFE CYCLE

Worms can live less than a year if their environment is not ideal. However, *Eisenia fetida* (striped worms) can live as long as 4 years.

Worms are hermaphroditic but mating is still necessary. Worms mate anywhere in the box and at any time of year if moisture and temperature conditions are right.

Before mating, a part of the worm called the clitellum, located about ⅓ of the way down the body, will swell to form a cocoon filled with eggs. A worm's sex organs are very close to the clitellum.

Worms mate by lying side-by-side with their heads in opposite directions, so that the sex organs line up with the clitellum. Sperm from each worm moves down a groove into the receiving pouches of the other worm.

After the worms have separated, the clitellum secretes a substance called albumin. This material forms the cocoon in which the eggs are fertilized and baby worms hatch.

Redworm cocoons are small and round. They change color during their development: white, then yellow, then brown, then finally, when new worms are ready to emerge, red.

Incubation in the cocoon is between 32 and 73 days. Temperature and other conditions affect the development of the hatchlings. Although a cocoon might hold as many as 20 eggs, usually only 3 or 4 worms will emerge.

The young hatchlings are whitish with a pink tinge,

which shows their blood vessels.

In about 8–10 weeks the baby worms will be mature and can begin reproducing. If conditions are good, a mature worm can make 2–3 cocoons per week for 6–12 months.

# Humanure for the City Dweller

As an urban gardener and forager living in Chicago, I am obsessed with improving soils. Composting is one major method of doing this (along with mulching and growing cover crops and mushrooms). Compost helps soil absorb and retain moisture and helps plants build nice, strong, and nutritious bodies upon which we may feed.

A few years ago, I started tentatively composting my body's own waste. It worked beautifully, and I have since completely abandoned my "flushie" and fully embraced my body as soil-maker (my toilet now serves as a bookshelf and plant stand).

Before indoor plumbing became common, human waste was a valued agricultural resource. "Honey wagons" moved waste from the city's bin commodes back into the country, where it was spread into fields to dessicate, solarize, and then be turned in and allowed to compost. It's a sustainable system that still operates in some places. Proper composting kills pathogens and creates lovely soil.

## Collect It

Your first step is basically to start pooping into a bucket toilet. These toilets are commonly used by campers, boaters, and hunters, and unlike pit latrines, they prevent the leaching of raw waste into soil and waterways. Bucket toilets concentrate waste so it can be of use later as *humanure* — a term popularized by Joseph Jenkins in *The Humanure Handbook*. The materials for such a toilet are entirely forageable, and its portability makes it easy to hide from squeamish visitors.

Bucket toilets are perfect for city dwellers, but be aware that using them and gardening with humanure are illegal according to most U.S. city health codes. So you should also hide your bucket from any visiting official.

All you need for your toilet is a common 5-gallon plastic bucket, obtainable from the back door or dumpster of most restaurants, and a carbon-rich cover material — usually sawdust, but you can also use newspapers, cardboard, dried leaves, straw, or composted stable bedding (call a local stable).

Sawdust and dry stable bedding are my favorite cover materials, as they both absorb moisture well. The sawdust I use comes from a furniture maker who uses hardwoods without glues or resins. You should be able to find a woodworker who would gladly let you sweep his or her floors for your resource.

I built a simple box for my bucket that stabilizes it under the weight of a body and hides its faded "Corn Syrup" label. The box has a hinged lid with a waxed top that's comfortable to sit on and cleans easily (Figure A). You can also find a snap-on toilet seat for a 5-gallon bucket online or at camping, hunting, or boating stores. Or just attach a regular toilet seat for a more familiar "toilet look."

To start your dry toilet, put a few inches of carbon material in the bottom. Then just use it the way you would use a regular toilet, and afterward cover your deposit well with your carbon material. Repeat as necessary until you have a full, compostable bucket.

Illustration by Alison Kendall

Photography by Nance Klehm (top left) and Ari Kletzky

## Compost It

Empty the contents of your bucket into the middle of your backyard compost pile (or in my case, into the large, lidded garbage bin tucked unobtrusively in the corner of my back porch). Cover with more dry carbon material to keep curious critters out and prevent it from losing too much moisture. Rinse your bucket with lightly soapy water (biosoluble or biodegradable soap), dump that into the compost, and the bucket can begin again.

The natural bacteria in your waste plus all the carbon and nitrogen in the compost pile will fire the mixture up to high temperatures pretty quickly. You're on your way to making soil from your soil!

If you don't have a backyard or storage area, you'll have to rely on friends and neighbors who do, or in the case of a few people I know, hide it in plain public space with signage declaring ownership ("Nance's Compost"). With all that wet sawdust, people won't know what they're looking at anyway.

## Dregs to Gold

Believe me, after a few seasons, if you've set things up reasonably well, your crap will have become lovely soil. If you remain unsure, and it has been a year, use the compost to nourish some trees or flowering plants (Figures B–D), or simply wait a few more months or a year to allow it to age.

If you're really nervous about trusting the natural thermophilic process to kill pathogens, test your soil through your state's land-grant universities (see the Membership Listing at www.aplu.org). The testing is low-cost or free, and can put your mind at ease.

## Troubleshooting

**Toilet is stinky, or has flies or maggots.** It is too wet. Use more cover material after each deposit.

**Compost pile is not heating up.** Your C:N (carbon to nitrogen) ratio is off. Add more nitrogen, in the form of urine, grass clippings (avoid atrazine herbicide), fruit and veggie peelings, etc. Your pile also may be dry, so add water and cover lightly with a tarp or some burlap to prevent evaporation.

**Compost pile is stinky.** It needs more oxygen. Incorporate dry material to create air space. If it's rainy, cover lightly with a tarp.

Whatever happens, just inhale deeply, and know that it's all going to be OK!

Nance Klehm is a radical ecologist, designer, and teacher who grows and forages most of her own food in a densely urban area. Her regular column "Weedeater" appears in *Arthur* magazine. spontaneousvegetation.net

# Solar-Heated Water

O ur quest to use solar energy started in 1996 when we decided to build a new home. In our area of Pennsylvania, the sun shines only about 50% of the time, on average, and the winters are cold. Contractors said that solar would not work, and the bank refused to lend us money for this type of home.

Fortunately, a good friend believed it could be done and financed our home so that we were able to build it as designed. Now that our heating bills are less than half those of our neighbors, we've proven the skeptics wrong.

For the past 12 years, our south-facing windows have saved us a lot on heating costs — but we still burned a fair amount of oil to make hot water. In spring 2008, when the price of heating oil was $4.50 a gallon, we searched for a way to use less oil. Our goal was to use none at all to heat our water during summer, while reducing consumption dramatically throughout the rest of the year.

Our system uses what I call the make-and-hoard method: it makes as much hot water as it can on sunny days, and stores it in well-insulated tanks for later use.

I started by purchasing two inexpensive 80-gallon stainless steel storage tanks and circulator pumps on eBay. These, along with our existing 40-gallon oil-fired tank, give us a total capacity of 200 gallons. Assuming the average person uses 20 gallons of hot water per day, the solar-heated system contributes sufficient hot water for my wife and myself for four days. And when it's full to capacity, the system continues to preheat our well water, reducing the amount of fuel oil we use to heat water in the 40-gallon tank.

Since the temperature in our area drops well below freezing during the winter, I used a closed-loop system with food-grade antifreeze. The solar collector is located at ground level on the hill behind my house, to allow me to keep the snow off and to maintain it when necessary (Figures A and B). It uses 90 Thermomax pyrex-glass vacuum tubes with coated copper strips inside each (Figures C and D).

The vacuum insulates these collector tubes for a

near-zero heat loss. The copper strips are attached to copper pipes that contain a small amount of antifreeze in a vacuum.

At 90°F, the antifreeze boils and the steam rises to the manifold, which heats water circulating through pipes that run underground between the solar collector and the heat exchangers in the tanks, located in our basement (Figure E).

A differential temperature-controller unit regulates the system. Sensors monitor water temperature in four locations: in the solar collector, in each of the two 80-gallon storage tanks, and in the common return line back to the collector (Figure F).

When the collector is warmer than the return line, the controller turns on the pump that circulates the heated water in the collector to the heat-exchanger

coil in the first tank. If there's enough sun, this circulation continues until the tank reaches 180°F, at which point the controller switches the flow to the second storage tank and heats the water in there.

If the second tank reaches 180°F, the controller will shut down flow to both tanks and turn on a small pump (Figure H, following page) to dump any extra heat to a radiator, which will safely dissipate the heat so that the system won't make steam. I also added a small bronze circulator pump to refresh the oil-heated tank with warm water from the first solar tank before the oil burner will run.

If the power fails, the collector will stagnate in the full sun, and high temperatures will build. Since many plastics melt under these conditions, I needed to find the right type of pipe to run from the collector to the

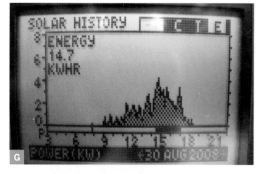

tanks and back, about 160 feet, round trip.

I was going to use PEX plastic pipe until I discovered EasyFlex, a newly approved, stainless steel corrugated pipe that's light, flexible, and perfect for solar hot water, at half the cost of copper. I ordered it with an exterior plastic coating, so that its corrugations won't act like a radiator and waste heat. The coating is rated to boiling temperature (100°C/212°F), and it made pulling 80 feet of pipe through a 4-inch PVC conduit much easier (Figures I and J). I recommend soldering (sweating) all the plumbing, because the temperature extremes can cause threaded pipe connections to fail.

The day we installed the tubes in the collector, the sky was overcast, with not enough sun to cast a shadow. Yet after installing only three of the 90

tubes, the system started to make warm water (Figure G) and store it in the basement tanks.

With the system fully installed, we've been able to generate more hot water than we can use. And it works great in the winter; even at −10°F outside, the system is making 160°F water to store, eliminating most of our need for fuel oil to make hot water.

Our cost for materials was $12,000, which will be reduced 30% with a federal tax credit. I estimate the system will pay for itself in six to seven years.

---

Dan Bassak is an electronics technician specializing in biomedical equipment repair and modification. He also builds gadgets that assist his wife, Trina, a blind physical therapist. They enjoy roller-skating, organic gardening, solar living, and bread baking.

# Sustainability Roundup

Volume 07, page 72

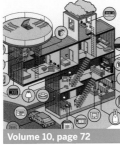
Volume 10, page 72

Volume 13, page 157

Volume 14, page 160

# Maker SHED
## DIY KITS + TOOLS + BOOKS + FUN

makershed.com

### PIR SENSOR
Detects motion from up to 20 feet away by using a Fresnel lens and an infrared-sensitive element. Ideal for motion-activated lighting, alarm systems, and robotics.

### XBEE ADAPTER KIT
With an XBee and our Xbee adapter kit, you can easily add wireless networking to your next project. Great for point-to-point, multi-point, and mesh networks.

### GRAVITY CLOCK
Build your own beautiful gravity clock and customize it using your own counter-weight. It's a perfect weekend project.

### SOLARSPEEDER KIT
Designed using gathered ideas from many years of Solaroller racing at the BEAM Robot games. Simple to construct and a great project for beginners!

### NEW! ARDUINO DUEMILANOVE
Now upgraded to a more powerful microcontroller: the ATmega328! Fully compatible with the previous ATmega168, but with twice the memory.

### SOLDER BUNDLE
Our Learn to Solder bundle includes 2 practice kits, MAKE, Volume 01, and our Maker's Notebook. It's a great bundle for anyone who wants to be a soldering pro.

# Make: Projects

DIY electronics make almost anything possible. With a single microcontroller you can make an indoor garden that reacts to temperature, light, and moisture. Or build an LED brick that gives fun, colorful, low-power illumination. Want to watch your energy bill? Use ready-made wireless modules to twitter your power usage to the web.

Photograph by Michael T. Carter

# THE GARDUINO GARDEN CONTROLLER

## By Luke Iseman

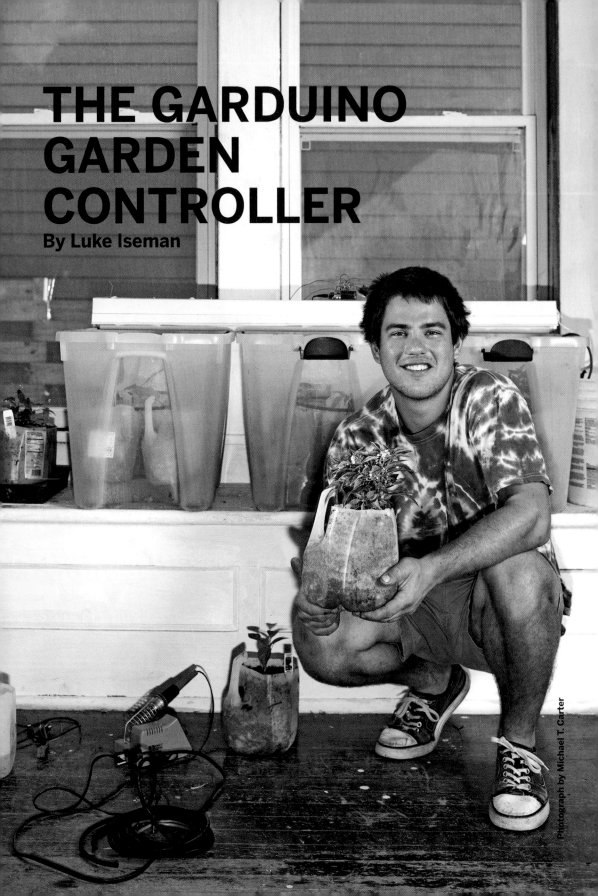

Photograph by Michael T. Carter

# GEEKED-OUT GARDENING

I wanted to start gardening, but I knew I wouldn't keep up the regular schedule of watering the plants and making sure they got enough light. So I recruited a microprocessor and a suite of sensors to help with these tasks.

My Garduino garden controller uses an Arduino micro-controller to run my indoor garden, watering the plants only when they're thirsty, turning on supplemental lights based on how much natural sunlight is received, and alerting me if the temperature drops below a plant-healthy level.

For sensors, the Garduino uses an inexpensive photocell (light), thermistor (temperature), and a pair of galvanized nails (moisture).

You can use a Garduino to experiment and learn what works best in your garden.

**Set up:** p.93   **Make it:** p.94   **Use it:** p.101

**Luke Iseman** builds open source hardware, including pedicabs, pallet furniture, and chicken tractors, in Austin, Texas. His projects are available at dirtnail.com

# MICROCONTROLLER-ASSISTED GARDENING

How the Arduino attends to your plants' every need.

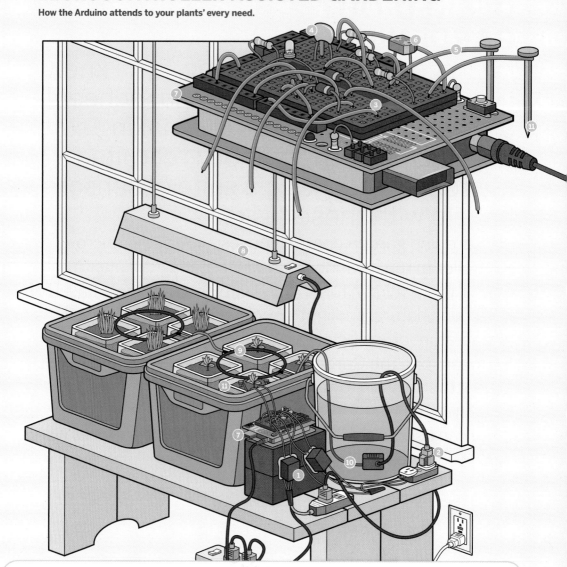

## INDOOR GARDEN

① Relays let you control AC power (lights and pumps) with the Arduino.

② Here we're controlling power to the pump with a relay wired to the Arduino.

③ The relays connect to our Arduino with diodes to prevent it from frying.

④ Thermistor: resistance changes with temperature.

⑤ These wires lead to nails stuck in the soil: our moisture sensor. Resistance between the nails, through the soil, changes with soil moisture.

⑥ Photocell: resistance changes with light intensity.

⑦ Garduino circuit board (Arduino + sensors).

⑧ The Garduino turns on the fluorescent light to make sure the plants receive the right amount of light (16 hours daily). The photocell plus Arduino measure how much natural sunlight is received and turn on the supplemental lights to make up any difference.

⑨ The black irrigation tubing has tiny holes that allow water to reach the plants when the pump is turned on.

⑩ The pump is submerged in a bucket of water.

⑪ Ordinary galvanized nails are used to sense soil moisture.

Illustration by Timmy Kucynda

# SET UP.

## MATERIALS

All electronics can be purchased at the Maker Shed (**makershed.com**) and Jameco (**jameco.com**), and everything else can be found at your local hardware store. See **makezine. com/18/garduino** for direct links to purchase the parts online. Total cost, including the Arduino, was about $150.

**[A]** Arduino microcontroller I used an Arduino Duemilanove, but any should work.

**[B]** Circuit board You can pack everything onto a ProtoShield (shown on previous page and available at the Maker Shed) but the following pages show a solderless breadboard for clarity.

**[C]** Omron G5LE-1 relays (2)

**[D]** 1N4004 diodes (2)

**[E]** 220Ω resistor for the LED

**[F]** Standard LED Any you'd use with an Arduino will do.

**[G]** Photocell

**[H]** 10kΩ thermistor

**[I]** 22-gauge wire, solid core and stranded, several feet

**[J]** 10kΩ resistors (3)

**[K]** Galvanized nails, 1"–4" long (2)

**[L]** USB cable

**[M]** AC extension cords (2)

**[NOT SHOWN]**

Plastic milk jugs for planting in. Use as many as you'd like. I used about 30.

Clear plastic storage containers, 28gal You'll need 1 for every 6 milk jugs.

Bricks or other spacers to raise the milk jugs at least 1" off the bottom of the bin. You'll need about 5 for each storage container.

Seeds preferably for things you'd like to eat. Swiss chard is an easy starter plant.

48" fluorescent light fixture

48" fluorescent tube "grow light" I used the OttLite, but any tube marketed for plant growth should be fine.

Soil mixture I used Mel's Mix, as recommended in Mel Bartholomew's *Square Foot Gardening*. It consists of 1/3 peat moss, 1/3 coarse vermiculite, and 1/3 mixed compost, with the mixed compost coming from at least 6 different sources. You can use whatever works for your plants.

Clean-water pump A small, cheap one is fine; I used a mini submersible pump from Harbor Freight (item #45303, available online at www.harborfreight.com).

Micro soaker hose kit also from Harbor Freight (item #65015). Or you can use bike inner tubes and poke holes.

5gal bucket

Funnel I used a cut milk jug

Photograph by Ed Troxell

**MAKE IT.**

# HOOK UP YOUR MICROCONTROLLER GARDEN

**START** ❖❖      **Time: A Weekend**    Complexity: **Moderate**

This project is ambitious for a first Arduino undertaking. I recommend completing at least the first few lessons of an Arduino tutorial before attempting this. There's a great one at ladyada.net/learn/arduino.

## 1. PLANT YOUR GARDEN

Sprout your seeds before planting them, or buy started plants. I planted a variety of vegetables in milk jugs with the tops cut off, with holes in the bottom to allow drainage, and a surrounding plastic storage container to catch water as it drained out.

## 2. MAKE MOISTURE SENSORS

**2a.** Cut 2 pieces of wire, each 2' long, and strip ½" off the ends.

**2b.** Wrap 1 end of each wire around the head of each nail.

**2c.** Cover the wire-nail connection with a generous amount of solder.

## 3. CONNECT MOISTURE SENSORS TO THE ARDUINO

You can tell when your soil needs water by measuring the resistance between the 2 nails stuck in the dirt. The more water in the soil, the more conductive it is.

**3a.** Connect a wire between ground on your Arduino and the ground (−) column on your breadboard. You'll use this column on the breadboard as ground for the rest of the circuit.

**3b.** Connect a wire between +5V on your Arduino and the positive (+) column on your breadboard. You'll use this column as the positive voltage connection for the rest of the circuit.

Photograph by Luke Iseman

**3c.** Connect one of the moisture sensors to +5V on the breadboard.

**3d.** Connect the other moisture sensor to a new row on the breadboard.

**3e.** Connect a 10kΩ resistor to the same row as the moisture sensor and also to a new row.

**3f.** Connect a wire from analog input 0 on your Arduino to the same row as the resistor and moisture sensor.

**3g.** Connect the other end of the resistor (in the new row) to ground.

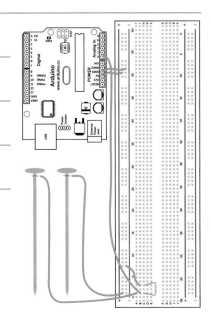

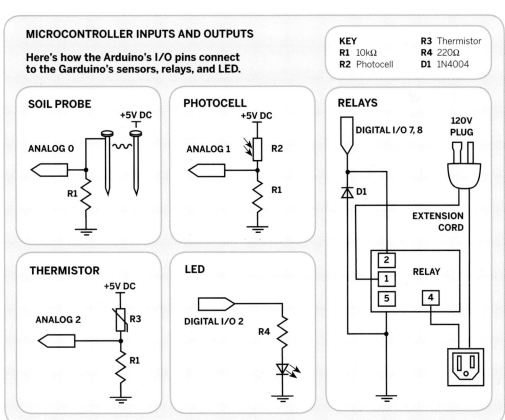

**MICROCONTROLLER INPUTS AND OUTPUTS**

**Here's how the Arduino's I/O pins connect to the Garduino's sensors, relays, and LED.**

KEY
R1  10kΩ
R2  Photocell
R3  Thermistor
R4  220Ω
D1  1N4004

SOIL PROBE
+5V DC
ANALOG 0
R1

PHOTOCELL
+5V DC
ANALOG 1
R2
R1

RELAYS
DIGITAL I/O 7, 8
120V PLUG
D1
EXTENSION CORD
2
1
5
RELAY
4

THERMISTOR
+5V DC
ANALOG 2
R3
R1

LED
DIGITAL I/O 2
R4

Illustrations by Gerry Arrington

# 4. LOAD THE SENSOR TEST CODE

Make sure that the moisture sensor works by connecting your Arduino to a computer and entering the sensor test code below (also at makezine.com/18/garduino). When you touch the nails together, the moisture value should read ~985; when they're not touching, the moisture value should be 0.

NOTE: If you haven't used your Arduino before, you need to connect it via its USB cable to your computer, then launch the Arduino development application (free download at **arduino.cc**), enter the code, and then upload it to the board. Select the serial monitor to see the output.

**CODE:**

```
int moistureSensor = 0;
int lightSensor = 1;
int tempSensor = 2;
int moisture_val;
int light_val;
int temp_val;

void setup() {
Serial.begin(9600); //open serial port
}

void loop() {
moisture_val = analogRead(moistureSensor); // read the value from the moisture-sensing probes
Serial.print("moisture sensor reads ");
Serial.println( moisture_val );
delay(500);
light_val = analogRead(lightSensor); // read the value from the photosensor
Serial.print("light sensor reads ");
Serial.println( light_val );
delay(500);
temp_val = analogRead(tempSensor); // read the value from the thermistor
Serial.print("temp sensor reads ");
Serial.println( temp_val );
delay(1000);

}
```

# 5. ADD THE LIGHT SENSOR

**5a.** First, connect the photocell to 2 new rows on the breadboard.

**5b.** Connect a wire between one row that the photocell touches and the positive column.

**5c.** Connect a 10kΩ resistor to the other row that the photocell touches and to a new row.

**5d.** Connect a wire between the photocell-resistor row and analog input 1 on your Arduino.

**5e.** Connect the other end of the resistor to ground.

**5f.** Test your light sensor by connecting your Arduino to your computer and monitoring the serial output. I measured the following values:
Indirect sun: 949
Ambient indoor light at night: 658
Ambient indoor light at night, with a
    hand casting shadow over the sensor: 343

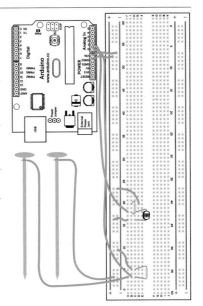

# 6. CONNECT THE TEMPERATURE SENSOR

**6a.** Connect the thermistor to 2 new rows on the breadboard.

**6b.** Connect a wire between one row that the thermistor touches and to the positive column.

**6c.** Connect the last of the 10kΩ resistors to the other row the thermistor touches and to a new row.

**6d.** Connect a wire between the thermistor-resistor row and analog input 2 on your Arduino.

**6e.** Connect the other end of the resistor to ground.

**6f.** Test your temperature sensor by connecting your Arduino to your computer and monitoring the serial output. I measured the following values:
61°F = 901
90°F = 949
51°F = 877
32°F = 796

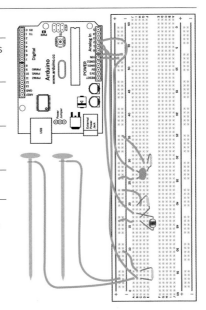

# 7. SPLICE RELAYS INTO THE LIGHT/PUMP POWER CORDS

Now we work with the heavy lifters: our relay setups. These will turn the lights and pumps on and off, in response to hours of sunlight received and soil moisture.

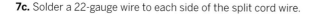

**7a.** Cut four 10" lengths of 22-gauge wire and strip ½" off each end.

**7b.** Look at your extension cord's plug: one prong is larger than the other. Split the 2 wires of the cord apart, then cut the wire that runs to the smaller prong, and strip 1" off each side.

TIP: **The correct wire is the one without ridges running along its length. Don't worry if you cut both wires; you can just splice the other wire back together.**

**7c.** Solder a 22-gauge wire to each side of the split cord wire.

**7d.** Solder the 22-gauge wire that runs to the extension cord's receptacle to the lower right lead of the relay (it should be labeled "4" on the bottom of the relay).

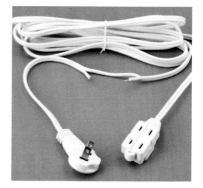

**7e.** Solder the 22-gauge wire that runs to the extension cord's plug to the middle left lead of the relay (labeled "1").

**7f.** Connect a 22-gauge wire to each of the 2 other leads on the left side of the relay (labeled "2" and "5"). Optionally, you can cover the relay's bottom side with hot glue to strengthen all 4 connections.

**7g.** Wrap both connections to the extension cord in electrical tape or heat-shrink tubing. Congratulations, you've completed your relay-cord setup.

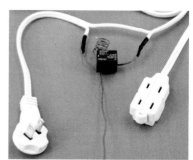

**7h.** Repeat Steps 7a–7g with another relay and extension cord to create the second relay-cord setup.

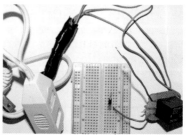

Photography by Ed Troxell; bottom photograph by Luke Iseman

# 8. CONNECT RELAYS AND AN LED TO THE ARDUINO

**8a.** Connect one of your diodes to 2 unused rows on the breadboard.

**8b.** Connect the bottom left lead (pin 2) of your relay (looking at it from the top, with the leads down) to the positive lead of your diode — the end that does not have a band on it.

**8c.** Connect the upper left lead (pin 5) of your relay to the negative lead of the diode — the end marked with a band.

**8d.** Connect a wire between ground on your Arduino and the ground column on your breadboard (if you're using the same breadboard for relays and sensors, just use the ground column you've already created).

**8e.** Connect the row containing the negative lead of your diode (the end with the band) and the upper left lead of your relay to digital input/output 7 on your Arduino.

**8f.** Connect the positive lead of the diode to your ground column. That's it for connecting the first relay.

**8g.** Now choose 2 new unused rows and repeat Steps 8a–8f to connect the second diode and relay as you did the first, except this one goes to digital input/output 8 on your Arduino.

**8h.** Connect the 220Ω resistor to 2 unused rows. Connect the LED's long leg (+) to either end of the resistor, and its short leg (−) to ground. Connect the other end of the resistor to digital input/output 2 on your Arduino.

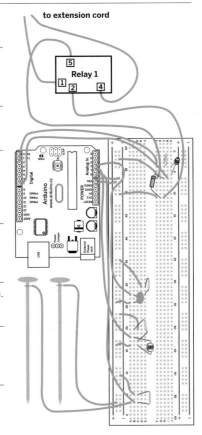

# 9. SET UP THE LIGHTING SYSTEM

An easy step here: after checking that your light fixture is working, plug it into the relay cord that will control it. I simply rested the fixture on top of the outer plastic bins, but feel free to get fancy. Optimum distance from these fluorescents for the light intensity you want is just a few inches, so make sure you get them up close, personal, and adjustable as the plants grow.

# 10. SET UP THE WATERING SYSTEM

You've got a wide variety of options here. Here's how mine works:

» A small pump is submerged in a 5gal bucket water source.
» A relay controlled by moisture sensors activates the pump to move water from the bucket to a milk-jug funnel.
» Gravity moves water from the funnel down to the soaker hoses, which drip into the plants.

I used a mini soaker hose kit from Harbor Freight Tools to assemble rings that drip into all the plant containers. If you make a setup like this, be sure to elevate the bucket on a crate or something else; moving water inches instead of feet vertically will greatly reduce the strain on your pump.

I initially tried using just a sprinkler valve mounted to the bottom of a bucket, without the pump. But gravity provided only enough pressure for the slightest trickle. I thought about connecting the sprinkler valve right into my plumbing, but I worried that the chlorine content of water straight from the faucet would be bad for my plants (chlorine evaporates from water within something like 24 hours).

A better version of this would be to use 2 buckets, with water coming from a sprinkler valve connected to house plumbing going into one bucket, being held there 24-plus hours, and then moving into the second, plant-feeding bucket.

# 11. PROGRAM YOUR GARDUINO

Lastly, you need to program your Garduino to run the garden. Because temperature and soil moisture are dealt with as constants (i.e., always turn on the LED if temperature is below a certain value, always turn on the water if moisture is below a certain value) they're simple to deal with.

Light is more complicated: you want to keep track of how much light your plants are getting, so that natural light plus supplemental light always equals optimum light time (in my case, I chose 14 hours daily). To do this, I used the *DateTime* Arduino library.

# FINISH

Photography by Michael T. Carter

# SWITCHING TO ...
# GARDENING COMPUTER

## REAP YOUR HARVEST

Check your seed packets (you saved them, right?) to see how many weeks until your plants should be ready for harvest. But don't be surprised if they're ready sooner than that! If they seem to be growing too slowly, check your watering and lighting routines.

## EVALUATE YOUR DATA

As currently implemented, Garduino needs to be hooked up to a computer that's monitoring serial output to obtain more meaningful data to share. With additional work, it's possible to store data on a USB drive. But for now, monitor the serial output in the Arduino environment to evaluate your Garduino's performance.

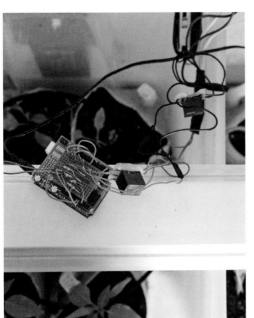

## IMPROVE AWAY!

I don't expect this beta Garduino to get everyone gardening and save the world; that's an exercise for readers to solve with their improvements. But here are some initial ideas:

» Use pulsing red and blue LEDs for an ultra-efficient lighting system (see screwdecaf.cx/sept.html for Mikey Sklar's version).

» Figure out what times of night your utility charges lower rates for electricity, and turn the lights on during those times only.

» Build a pH probe and fine-tune your soil acidity for different plants.

» Add a relay-controlled heater to keep a greenhouse version above a minimum desired temperature.

» Add a battery and solar panel to take the whole system off-grid.

» Use an irrigation valve instead of a pump to water your larger, outdoor garden, and add some modified solar garden lights for additional lighting.

If many people start recording the efficiency and convenience of this automated approach to gardening, then maybe we can even grow more food of better quality with less energy. Happy Garduino-ing!

➕ You can find the complete code at makezine/18/garduino. I'll add links to better versions as readers create them.

# LED LIGHT BRICK
## By Alden Hart

# A LITTLE GLOWING FRIEND

Restful, multicolored lights illuminate your path to the kitchen for a late-night snack. Rediscover the night light as an energy-efficient way to brighten your home.

The Little Glowing Friend (I named it after the ever-vigilant blue canary night light in a They Might Be Giants song) uses a single-chip microcontroller to drive 20 LEDs in a variety of patterns. The circuit is embedded in a clear resin casting, creating a lively, self-contained display. The light brick takes about 2 watts of power, and should last more than 10 years in continuous operation.

I gave away 24 of these in 2006; most have been on ever since, and there have been no reported failures. Electricity cost is about $1 a year, for 24×7 operation.

Set up: p.105    Make it: p.106    Use it: p.111

Alden Hart is CTO of Ten Mile Square Technologies, a technology consulting firm that develops systems for media and communications, from the metadata to the metal. In his spare time he combines micro-controllers, LEDs, mechanics, and other small parts in ways that have no practical application.

# CHIP AHOY

The heart of the Little Glowing Friend is a single-chip microcontroller (PIC16F916) that's programmed using PIC assembly code.

This chip provides the following functions:
» Drives the LEDs directly from the chip's output pins
» Provides 20 dimming channels — one for each LED
» Stores and runs the light patterns

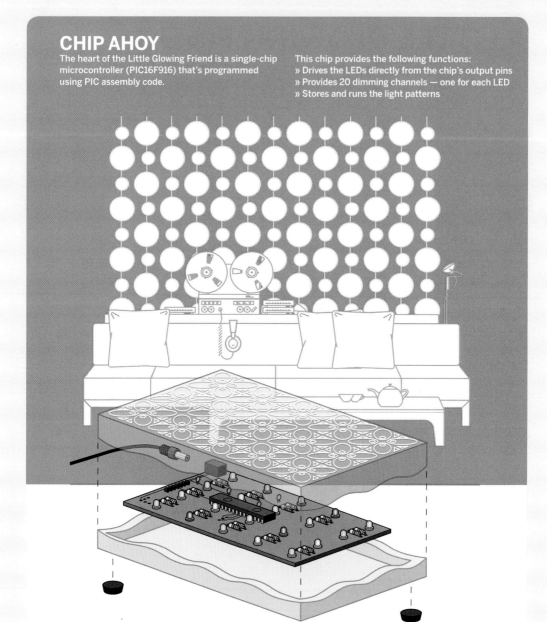

The printed circuit board (PCB) holds the PIC microcontroller, LEDs, current-limiting resistors, a DC power connector, and a few other parts.

The PIC chip is capable of driving up to 25 milliamps (mA) of current per output pin to the LEDs. A current-limiting resistor is used to limit the current flowing through each LED to about 20mA. There are 2 decoupling capacitors that clean up any noise that might be on the power supply. Finally, one resistor is used to hold up the PIC's reset line so the chip can run, and another is used for the switch.

Because LEDs glow at only one intensity, dimming is accomplished through a variant of pulse width modulation (PWM) — flashing the LED on and off so fast that your eye sees it as a brightness level less than 100%.

Once the circuit is working, the entire board is embedded in a clear resin casting that lights up in colorful and interesting ways. You can make your own mold as part of the embedding process, so the brick can take on any shape and texture you desire.

Illustration by Nik Schulz

# SET UP.

## MATERIALS

**[A]** Circuit board from the light brick kit or make your own from the CAD files. Both are available at makezine.com/18/lightbrick.

**[B]** PIC16F916 microcontroller labeled as U1 on the PCB and the schematic, which is also available at the link above

**[C]** LEDs, 5mm, 20mA, wide viewing angle, 0.100" lead spacing (20) 5 each, red, yellow, green, and blue, D1–D20 on the PCB and schematic

**[D]** 100Ω resistors, ¼ watt, 5% carbon film (20) R1–R20

**[E]** 20kΩ resistors, ¼ watt, 5% carbon film (2) R21, R22

**[F]** 10µF capacitor, tantalum, 16 volts, 0.1" lead spacing C1

**[G]** 0.1µF capacitor, monolithic, 0.100" lead spacing C2

**[H]** Rolling ball tilt switch, 0.100" lead spacing SW1

**[I]** DC power connector, 2.1mm ID coaxial J1

**[J]** Rubber bumpers or feet (4)

**[K]** DC power supply (wall wart), 5V DC regulated, 300mA or more, 2.1mm ID/5.5mm OD, center positive I recommend a 400mA supply because sometimes 300mA ratings are a bit optimistic.

**[L]** 12" length of #14 solid wire (optional)

**[M]** 22pF capacitors, monolithic or ceramic disc, 0.100" lead spacing (2, optional) C3, C4

**[N]** Single-row header connector, 6-pin, 0.100" (ICSP connector)

*For molding, use O and P to make your own mold, or use Q instead.*

**[O]** Silicone RTV Mold-Making System, 1lb silicone and 0.1lb catalyst from TAP Plastics (tapplastics.com). My mold used all 16oz; if you design a bigger mold, get more.

**[P]** Lego bricks for building a mold box

**[Q]** Self-releasing polypropylene mold, 3"×5"×1¹⁄₁₆" such as Castin' Craft MC-7, part #43893

**[R]** Clear-Lite polyester casting resin, 16oz from TAP Plastics

**[S]** MEKP liquid catalyst, ½oz

**[T]** Paper cups, 12oz or 16oz (2) for measuring and pouring

**[U]** Stirring sticks (2)

**[V]** Clear gloss resin spray (optional) such as Castin' Craft Resin Spray, for covering imperfections

NOTE: A complete bill of materials and tools, with sources, as well as mold-making and resin-casting instructions, are available at makezine.com/18/lightbrick.

Photography by Alden Hart

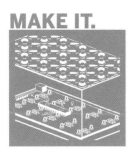

## MAKE IT.

# BUILD YOUR LED LIGHT BRICK

**START** »»        Time: **A Weekend**  Complexity: **Easy to Difficult**

Building the light brick involves 3 steps: assemble the PCB, build (or buy) the mold, and cast the piece. The project can be done over a weekend, with each step taking a few hours. Some steps require a wait time, so the total build requires at least 2 days and an overnight to cure the casting. The project is designed at multiple levels of difficulty:

**Easy:** Assemble the circuit board from a kit and do the acrylic embedding with a pre-made mold.

**Medium:** Assemble the circuit board from a kit. Then do the mold construction and casting.

**Difficult:** Program the PIC yourself, to make new patterns. If you want to get really advanced, you can lay out and make your own circuit board — allowing you to change the size and layout and even the number of LEDs. You could also "free-space" the wiring. All the source materials to do this are available online.

For kits, CAD files, firmware, and additional documentation, visit makezine.com/18/lightbrick, or my website at tenmilesquare.com/light-brick.

# 1. ASSEMBLE THE CIRCUIT BOARD

First, locate the top and bottom of the PCB. The top has the silk-screen and component designators. The bottom has most of the wiring.

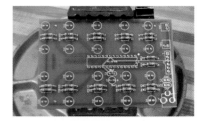

**1a.** Solder the resistors first. Use 100Ω resistors for R1–R20, and 20K resistors for R21 and R22. Resistors are not polarized, so it's not necessary to line up the color codes, but it does make for a neater job. Tape the resistors tightly to the top of the board using blue masking tape; they'll trap air bubbles underneath them if they're not down tight.

Flip the board over and solder. If any of the resistors shifted, heat them up and push them back down. Cut the leads down to the solder joints with wire cutters and remove the blue tape.

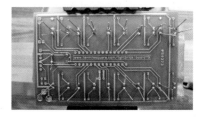

**1b.** Place capacitors C1 and C2 on the board. The tantalum capacitor C1 is polarized and must be inserted as indicated on the board. The positive (+) side of the capacitor goes in the + hole, which is the square one. Tape, solder, and cut as before.

**1c.** Carefully bend the leads of the PIC to fit the board's chip hole spacing. Place the PIC with pin 1 (the indented end) pointing to R21 and R22. Carefully solder, without putting too much heat on any pin. Double-check that the polarity is correct before soldering.

**1d.** Attach the DC power connector J1. Bend the blades out somewhat from the bottom to hold the connector in place. Then solder it in flush to the surface of the board, using plenty of solder to seal the holes.

**1e.** Attach LEDs D1–D20. The patterns are programmed for the following layout of LEDs:

**R B G Y R**
**Y R B G Y**
**G Y R B G**
**B G Y R B**

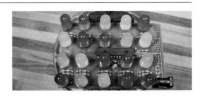

Be sure to test your LEDs for function, and for color if it's not obvious by the casings. Use an LED tester or just squeeze the leads onto a 3V lithium coin cell.

For proper casting, you want the tops of all the LEDs to be roughly level, and taller than the power connector. The holes in the PCB are sized so that a standard LED lead will stop about ⅛" above the board. It's a tight fit, but it works.

If for some reason your LEDs don't have this stop, you'll need to make spacers. Make 20 spacers by stripping ⅜" of insulation from #14 solid wire. Insert the insulation between each LED's leads, up by the LED case.

Place the LEDs in the board and test the heights against the power connector. Observe correct polarity when placing the LEDs. The long lead is the positive, or anode (A), and goes through the round hole. The short lead is the negative lead, or cathode (K), and goes through the square hole (mnemonic: "cats are negative"). If there's a flat spot on the LED it will be on the cathode side.

NOTE: The square hole here is designated differently from the capacitor's, which is the opposite.

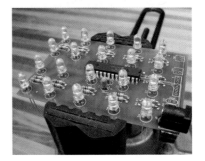

**1f.** Turn the board upside down on a hard, flat surface to ensure all LEDs are lined up. Solder, and clip the leads. Remove any spacers using hemostats or fine pliers. Notice that the common rail is connecting all the positive (+) pins of the LEDs. (In some other LED projects there is a common ground (−) for all the LEDs.)

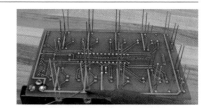

**1g.** Attach tilt switch SW1. The tilt switch is a rolling ball that runs on a track and closes (activates) when the ball rolls to the base where the wires stick out. It provides user input, acting like a button push or mouse click. The tilt switch is the only component that goes on the bottom of the board. It doesn't have a polarity. If you want it to activate when you tip the brick back toward the power connector, the switch must angle downward from the rear of the board, as pictured.

If you want this orientation, mount the switch on the bottom of the board through the switch holes labeled SW1. Leave about ¼" of the switch leads exposed between the switch and the board. You'll need enough room to bend the leads to angle the switch. Use the SW2 position if you'd rather the switch activate 90° from the SW1 position.

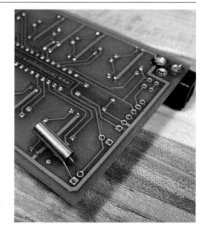

**1h.** If you intend to program the PIC in-circuit, you'll need to attach capacitors C3 and C4 and the programming header J2 at this point.

**1i.** Don't plug the board in just yet. Visually inspect the PIC and the rest of the board for solder bridges (i.e., short circuits in your soldering). Remove any bridges with solder wick (for advice on how to do this, see makezine.com/18/lightbrick).

Next, test the board for a short circuit. Use a continuity tester or ohmmeter to test the power connector's + and − terminals (located on the bottom of the board).

Finally, test that you have a good power supply. The center pin on the power supply connector should be a stable +5V, and the sleeve is ground. Be sure not to short out these contacts during this test! The board will actually work between about 4V–6V. The PIC doesn't like voltages above 6V and may blow out if overvoltage is applied.

If the board passes these tests, then plug it in! If the board doesn't work properly, consult the Troubleshooting section on page 111. Don't cast the brick until the circuit is working reliably.

# 2. MAKE THE MOLD AND CAST THE BRICK

Since we've covered moldmaking and casting in a previous issue (see *MAKE, Volume 08, page 160*), we won't detail the process here. Complete moldmaking and resin casting instructions for this project are available at makezine.com/18/lightbrick. The easiest way to get a mold is to buy a pre-made one like the MC-7 (see Materials list). It's 3"×5"×1⅟₁₆", a reasonable fit.

I prefer my own mold, for a better fit and a more interesting finished piece. The circuit board is designed to fit nicely into a volume 3"×4¼"×1"–2½" deep, but you can make the master mold any size and shape that will accommodate your board. Remember that the power connector needs to be flush against one side of the finished casting.

You can also house the Little Glowing Friend circuit board in any container you desire, but I like the permanence and uniqueness of the cast brick.

# 3. FINISHING UP

**3a.** Don't get impatient and demold the piece too early, as this can ruin it. Follow the manufacturer's recommended demold times, and then some. It's best to leave it overnight or even longer.

Remove the brick when it's truly cured. It should be fully cooled and hard. Resin hardens from the inside out, so the surface is the last part to harden. This can take well over a day depending on the mix and conditions. Don't judge by time; demold only when the surface is hard and no longer tacky. Test hardness using a stick, not your finger.

You can speed the surface cure by warming the brick under some lights, but be careful not to overheat the casting or the mold. Don't leave the lights on overnight or unattended.

**3b.** Remove the tape from the power connector using hemostats. You may also need a knife if it's gotten coated over.

**3c.** Even though the brick is hard at this point, the finish is still fragile. It will pick up fingerprints and will pit with dust. It's best to handle it only by the edges. You may want to "tent" it under wax paper and continue the cure. Don't let the wax paper touch the surface, or it will leave marks. Optionally, you can spray on a surface coat of resin at this point to protect the finish, but be aware that this may cloud the surface. Use sparingly.

**3d.** Apply the bumper feet to the bottom and you're done.

**FINISH** ☒

**NOW GO USE IT »**

# MEET PEGGY, OUR COVER SPOKESLIGHT
## By Gareth Branwyn

Geek super-couple Windell Oskay and Lenore Edman like to trip the light (emitting diode) fantastic. One of their first projects to blip the global DIY radar was their interactive LED dining table, with 448 LEDs under frosted glass.

After the bizarre Mooninite Invasion of 2007, when some Lite-Brite-type signs got the better of Boston's finest, Oskay and Edman, who invent under the name Evil Mad Scientist Laboratories (evilmadscientist.com), were inspired to create Peggy, a plug-and-play (or plug-and-program) LED "pegboard" you build from a kit.

With the success of Peggy, the white coats at EMS Labs knew they were on to something. "It became clear that there was room for something more advanced," says Oskay. "So we made the Peggy 2, which has the same basic design but supports simple animations."

Recognizing the strength of the Arduino community, they designed Peggy 2 to be programmed through the Arduino environment. "The architecture is similar to that of an Arduino clone, but with substantial onboard LED driving hardware," he says.

So how long does it take to build a board with up to 625 LEDs? An hour or two on the rest of the board, and then the time to solder on the lights. You decide how many you want to install — the "resolution" of your screen. Only rudimentary soldering skills are required. To create a basic sign lighting every LED, no programming is needed. To animate Peggy, and create other lighting effects and control, programming can be done via Arduino.

"People love interacting with the Peggy," says Oskay. "There's something about giant pixels — especially giant physical pixels that you can touch — that really grabs people. We love seeing what folks do with it. One builder, Jay Clegg, even figured out how to display grayscale video on the Peggy 2."

The original Peggy is $80 and Peggy 2 is $95, LEDs not included. Both kits are available in the Maker Shed (makershed.com) and the Evil Mad Science Shop (evilmadscience.com).

---

Gareth Branwyn is senior editor at MAKE.

Photograph by Garry McLeod

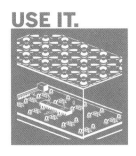

# FIRE UP YOUR LITTLE GLOWING FRIEND

Plug it in and watch it go! Activating the tilt switch will cycle the device between power off and 4 different lighting programs — Waves, Colors, Calm, and Frenetic.

## TROUBLESHOOTING

If, for some reason, the board doesn't light when you test it (in Step 1i, before casting the brick, of course), check the following:

» Ensure that all leads go through the holes, and check that none of the PIC leads have folded up under the chip body. They should all be visible from the bottom of the board.
» Inspect the bottom of the board to ensure that all connections are soldered.
» Check the polarization and orientation of components C1, U1, and D1–D20.
» Check that there are no solder bridges (two points connected that shouldn't be).
» Check that the power supply is plugged in and outputting between 4 and 6 volts. Use a volt/ohm meter (VOM) or multimeter.
» Check that the center pin of the power supply is positive, not negative (reversed polarity).
» Check that the board is not short-circuited by testing the + and − terminals on the bottom of the board.

If some LEDs are working but not others, try doing the following:

» Test the LED individually. Apply voltage from a coin cell across the LED terminals.
» Next, test with a VOM that there is 100 ohms of resistance between the LED's negative terminal (the square hole) and the PIC pin. Trace the circuit visually or use the schematic diagram.

## RESOURCES

➕ For a materials and tools list, sources, CAD files, a schematic, and moldmaking and resin-casting instructions, visit makezine.com/18/lightbrick.

Connecting LEDs: makezine.com/go/led2
Free-space wiring example: instructables.com/id/3x3x3-LED-Cube

🎥 Video on how to remove solder bridges with solder wick (see 7:45): makezine.com/go/solder

*Thanks to Daniel Klaussen for helping us in testing this project.*

# TWEET-A-WATT
# POWER MONITOR

**By Limor Fried and Phillip Torrone**

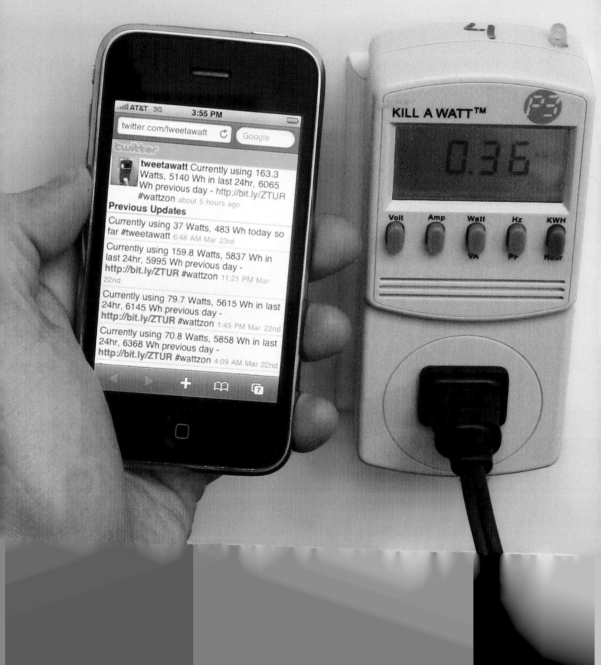

# WATT WATCHER

We live in a rented apartment, so we don't have hacking access to a power meter or breaker panel. But we still wanted to measure our household power usage long-term, so we developed the Tweet-a-Watt. It uses plug-in electricity monitors at each outlet to wirelessly send readings to a base station, which assembles them into reports you can analyze and graph. It can also broadcast updates via Twitter.

Building your own power monitor isn't too tough and can save money, but we're not fans of sticking our fingers into 120V wiring. Instead, we built on top of a P3 Kill A Watt power monitor, which we found at the local hardware store.

To track usage room by room, for example "kitchen," "bedroom," "workbench," and "office," you can use a 6-outlet power strip in each room to feed all the room's devices through a shared monitor. Each Kill A Watt can measure up to 15 amps, or about 1,800 watts, which is plenty for any normal room.

You can build each wireless outlet monitor for about $50 with a few easily available electronic parts and light soldering, and about the same for the receiver. No microcontroller programming or high voltage engineering is necessary!

**Set up: p.115    Make it: p.116    Use it: p.122**

**Limor Fried** is owner and operator of Adafruit Industries (adafruit.com), an open source hardware electronics company based in New York City. **Phillip Torrone** is senior editor of MAKE magazine.

Photography by Limor Fried

# TWEET-A-WATT: MAKING OUTLETS TALK

In this system, each Kill A Watt monitor transmits power usage data
to the base station computer via XBee wireless modules.

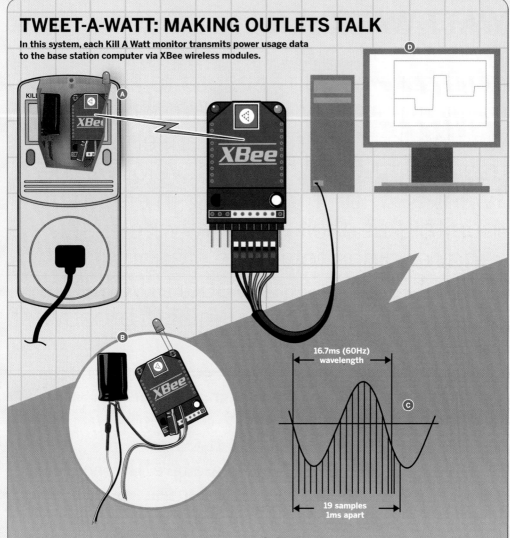

## Hardware

Ⓐ The XBee radio transmitter fits inside the Kill A Watt's case and taps into its amplifier chip to draw analog readings for voltage and current. Multiplying these values together gives you power in watts ($P = VI$).

For its power, the XBee draws from the Kill A Watt's power supply. A voltage regulator circuit reduces the Kill A Watt's internal 5V to the 3.3V needed for the XBee. The sensors have voltage dividers to do the same.

Ⓑ Because the Kill A Watt can only spare about 1mA of current without being disrupted, 2 capacitors smooth out the XBee's power demands. A big one (220µF) buffers power for the XBee's periodic startup, and an enormous "supercap" (10,000µF) buffers power for its radio transmission bursts.

## Software

Ⓒ Since we're dealing with AC, the voltages follow a 60Hz sinusoidal curve. Our XBee samples this curve in a quick burst every 2 seconds, with each burst containing 19 instantaneous samples spaced 1 millisecond apart. The individual sample numbers follow the AC sinusoid up and down, and the code averages these to derive a single number reading for each burst, a "snapshot" of one sine-cycle. These averaged readings still scatter a bit, depending on where each 18ms-long sample burst happens to land on top of the continuous 60Hz AC wave. But averaged over time, the slight variation washes out.

Ⓓ Tweet-a-Watt runs on the base station computer, where it shows the volts and watts consumed and records them to a logfile. You can set options in the code to make it tweet the reports (send them as Twitter messages) and graph your power usage history.

Illustration by Tim Lillis

# SET UP.

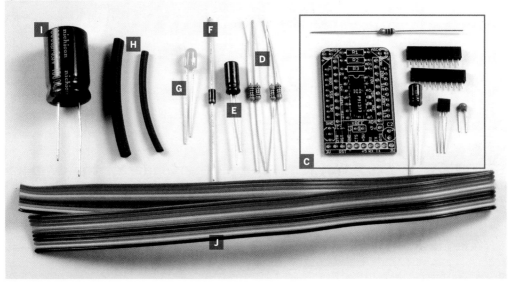

## MATERIALS

**FOR EACH OUTLET YOU WANT TO MONITOR:**

**[A] Kill A Watt, Model P4400** about $25 from a hardware store

**[B] XBee Series 1 RF module with flat chip antenna** XB24-ACI-001, about $20 from the Maker Shed (makershed.com), Adafruit (adafruit.com), Digi-Key, or Mouser

**[C] XBee adapter kit** $10 from Adafruit, or you can use another breakout/carrier board with a regulated 3.3V power supply and status LEDs

**[D] Resistors, ¼W, 5%:** 4.7kΩ (2), 10kΩ (2)

**[E] 220µF capacitor, 4V or higher** Try to get 5mm diameter.

**[F] 1N4001 diode** Digi-Key or Mouser

**[G] Large diffused LED (optional)** for easy viewing

**[H] Heat-shrink tubing, ¹⁄₁₆" and ¹⁄₈" diameter** about 2" each

**[I] 10,000µF capacitor ("supercap"), 6.3V** Try to get 16mm diameter.

**[J] Ribbon cable, 4-contact, about 32"** Rainbow coloring makes it easier, or use flexible wire.

**[NOT SHOWN]**

**Double-sided foam tape or hot glue**

**FOR THE BASE STATION [NOT SHOWN]:**

**FTDI cable (TTL-232R), 3.3V or 5V** $20 from Adafruit, connects XBee 6-pin serial adapter to computer USB

**Access to Windows-based computer** for updating XBee Firmware. You can run the Tweet-a-Watt code on any sort of computer; you only need a Windows machine to build it.

## TOOLS

**Soldering iron** preferably with temperature control, stand, and conical or small flat tip

**Solder, 60/40 rosin core**

**Desoldering tool**

**Flush/diagonal cutters**

**Helping Hands tool with magnifying glass (optional)** makes things go much faster

**Dremel rotary tool or small drill**

**Small screwdriver** to open the Kill A Watt case

**Heat gun or hair dryer** for heat-shrink tubing

# MAKE IT.

# BUILD YOUR TWEET-A-WATT

**START** ❯❯    Time: **A Weekend**  Complexity: **Moderate to Advanced**

# 1. PREPARE THE WIRELESS MODULES

**1a.** Assemble one of the XBee adapter kits into an XBee breakout board, following the included instructions.

**1b.** Plug an XBee module onto the XBee breakout board and connect it to your computer via the FTDI cable. Plug the cable into the 6 XBee pins running from GND (ground) to CTS (flow control), marked by a strip of metal flashing.

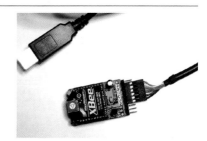

**1c.** Identify which serial port (COM) the cable is connected to. Under Windows, open the Device Manager and check how the port is listed. Ours was COM4.

**1d.** Download and install X-CTU software, which you can find by searching for "X-CTU" at digi.com. Launch the software, which configures and tests the XBee's radio.

**1e.** In X-CTU, under the PC Settings tab, select the port you identified in Step 1c and set the connection properties to 9,600bps, 8 bit, no parity, 1 stop bit, no flow control (or 96008N1). Click the Test/Query button. A pop-up window should tell you that the communication was OK.

**1f.** Under the Modem Configuration tab, click the Read button to read in the current version and settings of the firmware. Download the latest version by clicking the Download New Versions button and selecting Web. Then select the last version in the Version dropdown on the right, and click the Write button to upload it to the XBee. Check the Always Update Firmware checkbox to keep things up-to-date.

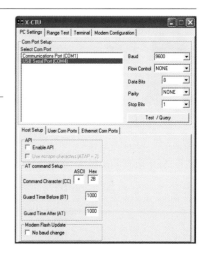

**1g.** Unplug the XBee module and repeat the upgrade for your other modules. You can hot-swap if you're careful: unplug the adapter from the cable first, gently swap modules, and plug in the adapter again. Then run Test/Query, Read, and Write from X-CTU.

**1h.** Now we'll configure the monitor modules to send the data we want and sleep between transmissions. For each XBee module (except for one), plug it back into the cable, run X-CTU, and under Modem Configuration click Read to load the settings. Configure the modules as follows, scrolling down as needed to click and enter parameters in place:

» Set the MY address (the identifier for the XBee) to 1, then increment for each additional transmitter to tell them apart.
» Set the Sleep Mode SM to 4 (cyclic).
» Set the Sleep Time ST to 3 (3 milliseconds after wake-up, go back to sleep).
» Set the Sleep Period SP to C8 (0xC8 in hexadecimal = 200 = 2 seconds between transmits).
» Set ADC 4 D4 to 2 (enable analog-digital converter D4).
» Set ADC 0 D0 to 2 (enable analog-digital converter D0).
» Set Samples to TX IT to 13 (0x13 = 19 samples).
» Set Sample Rate IR to 1 (1ms between samples).

Basically this means we'll have a single PAN network, each XBee will have a unique identifier, they'll stay in sleep mode most of the time, and they'll wake up every 2 seconds to take 19 samples from ADC 0 and 4, 1ms apart.

**1i.** Click the Write button to upload the new settings. You should see the green activity (RSSI) LED blink every 2 seconds, indicating wake-up. Note that once the XBee is told to go into sleep mode, it won't communicate with X-CTU until you reset it by unplugging it from the FTDI cable. Repeat Steps 1h–1i for the other transmitter modules. Then label the XBees, using a Sharpie, stickers, or similar, so you can tell which ones are transmitters and which one is the receiver.

# 2. ASSEMBLE THE TRANSMITTER BOARDS

For wiring, refer to the schematic diagram at makezine.com/18/tweetawatt.

**2a.** The unassembled XBee adapter kit(s) are for the transmitter modules. With each board, solder in the power supply components (labeled C1, C2, and IC1 on the schematic), the sockets, and the LED marked ASC. Don't install IC2, R1, R3, or the RSSI LED. Don't clip the legs of the ASC LED, so that it will extend out to fit through the Kill A Watt case, and you can optionally replace it with a brighter LED.

**2b.** To set the analog sensor's voltage reference to 3.3V maximum, solder a wire between the VCC pin (top left) and the VREF pin (lower right).

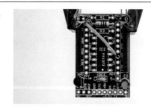

**2c.** Follow the schematic to solder in the two 10K resistors R4 and R6 from the XBee's AD0 and AD4 (analog ins) to ground. Since we're not installing the voltage level shifter IC2, just run the resistors to the closest grounds, the pads for IC2 pins 10 and 13. These resistors form a voltage divider that reduces the Kill A Watt's 5V signal down to 3.3V for the XBee.

**2d.** Cut and peel a 6", 4-wire length of ribbon cable. Solder 2 adjacent strands to the 4.7K resistors, reinforcing with heat-shrink tubing. These wires will carry the current and voltage readings from the Kill A Watt. Clip the other ends of the 4.7K resistors short and solder them to piggyback on top of the 10K resistors on the analog-in side.

**2e.** Solder the 2 other wires from the ribbon to +5V and GND on the breakout pin contacts at the bottom of the board.

**2f.** Solder in the reset capacitor, C3, between the XBee's reset pin (RST) and the nearest ground (the pad for IC2, pin 4). The long lead side (+) connects to reset. Give it some lead length so you can bend the cylinder down to tuck it next to the 3.3V regulator (IC1). This cap trickle-charges on the reset line so that the XBee waits a few seconds to start up. This prevents the XBee from drawing too much current from the Kill A Watt.

**2g.** At the other end of the 4-wire ribbon, solder the +5V and GND wires to the positive (+) and striped negative (−) terminals of the enormous capacitor C4. This capacitor continuously vampires power from the Kill A Watt, then dumps it to the XBee every few seconds when the radio transmits.

**2h.** Trim the legs of the supercap and solder its anode (+) leg to diode D3's striped cathode (−) end. This diode makes doubly sure that the capacitor won't be drained by the Kill A Watt.

**2i.** Solder 2 more wires to the ends of the supercap's legs, black to − and white or red to +. Then heat-shrink over the whole diode and the exposed capacitor leads.

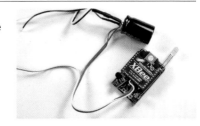

# 3. ASSEMBLE THE SENSOR/TRANSMITTERS

Now, the fun part — we'll fillet, stuff, and reassemble the Kill A Watt!

**3a.** Open the Kill A Watt by removing the 3 screws. Be careful not to damage the ribbon cable holding the sides together!

**NOTE: If your Kill A Watt looks a little different from this photo, that's OK. The innards are the same. You may just need longer wires to place the supercap.**

**3b.** Now it's time to jack into the sensor output! Melt a bit of solder to "tin" the ends of your power, ground, and sensor wires from the XBee board, then tin pins 1, 4, 11, and 14 of the LM2902 op-amp chip inside the Kill A Watt. Connect power to pin 4, ground to pin 11, analog in AD0 to pin 14, and AD4 to pin 1.

**3c.** Plug a configured XBee module into the adapter board, put some foam sticky tape or hot glue on the back, and fit it into the Kill A Watt case. Also find a place for the supercap, using tape or glue if needed. They should fit with no problem.

**3d.** Find a good place in the Kill A Watt case for the activity indicator (RSSI) LED, and drill a hole to fit. This lets you see when the XBee is transmitting.

**3e.** Close it up and plug it in. You'll notice it's a bit finicky for a few seconds as the big capacitor charges up, but after about 15 seconds you should see the display stabilize and the red LED blink every 2 seconds.

# 4. PREPARE THE BASE STATION

**4a.** Go back to your computer and plug the receiver XBee into the USB adapter. You should see its RSSI LED light up whenever the transmitters send data. That means you have a good link!

**4b.** Run X-CTU and open the Terminal tab. You'll see a lot of junk, but what's important is that a new chunk is added every 2 seconds. The hardware is done. Good work!

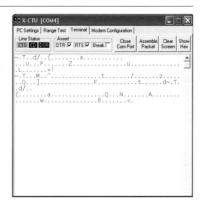

# 5. INSTALL AND CONFIGURE THE SOFTWARE

**5a.** If you don't already have Python, download and install it from python.org/download. We built Tweet-a-Watt using version 2.5, so that's the safest. We installed it into the directory (file folder) *C:\python25*.

**5b.** Download and unzip the Tweet-a-Watt code bundle from makezine.com/18/tweetawatt into a new project directory, *C:\wattcher*. This includes the *wattcher.py* Python script. Download 3 additional libraries linked from the same page, and put them into this project directory: *win32file*, so Python can write out the power data log file; *pyserial*, so it can communicate with the XBee through the serial port; and *simplejson*, to call the Twitter API.

**5c.** Open *wattcher.py* in a text editor and change the **SERIALPORT** = line near the top (line 24) to specify your serial port, from Step 1c.

**5d.** Make sure one of your sensor/transmitter units (Kill A Watt + XBee) is plugged in and blinking every 2 seconds, and that the receiver XBee's RSSI LED is also blinking. Nothing should be plugged into the Kill A Watt yet, and its LCD should be clear, not fuzzy.

**5e.** Open a Terminal window (**Run cmd**), navigate to the project directory (**cd c:\wattcher**), and run Python on the *wattcher.py* script (**C:\python25\python.exe wattcher.py**). You should see a steady stream of data listing current, wattage, and watt-hours since the previous reading. Hooray! We have wireless data!

**5f.** The output of the script probably says that power is being used, so we need to calibrate the sensor in the code so that it knows where zero is. Type Ctrl-C to quit the script, then run it again with the debug flag -d (**C:\python25\python.exe wattcher.py -d**). In the output, look for lines that start with **ampdata** and note the number that follows. Quit the script, open *wattcher.py* in an editor again, and change the **VREF** calibration number listed for the sensor to this number.

**5g.** Repeat Steps 5d–5f to test and calibrate your other sensor/transmitters one at a time.

**5h.** Run the code, and watch the watts! A nice way to test is by sticking the meter on a dimmable switch and seeing how the dimmer affects the numbers.

**5i.** Finally, let's configure the Tweet-a-Watt to tweet. If you don't already have a Twitter account, sign up at twitter.com/signup. Then edit *wattcher.py* again, and change the 2 lines defining **twitterusername** and **twitterpassword** to specify your username and password. Then run the script as usual, and it will send a Current Usage tweet every 8 hours, at midnight, 8 a.m., and 4 p.m.

# FINISH ✕

**NOW GO USE IT »**

# USE IT.

# TOTAL POWER AWARENESS

## GRAPHING THE DATA

An extra feature built into the Python script is code that graphically displays 2 hours of energy usage coming in from one of the sensors. To run this, install into the project directory the free graphing modules wxPython (wxpython.org), matplotlib (matplotlib.sourceforge.net), and NumPy and PyLab (both from scipy.org). Then edit *wattcher.py* to set GRAPHIT = True.

## BASE STATION VARIATIONS

Because the web/Twitter app uses Google authentication, we can feed data into it from a computer. A great enhancement would be to program an Arduino so that it could update the app's database, either by getting it to simulate a computer or by using another authentication scheme. The Arduino would let the base station be small and portable, drive outputs such as displays through simple direct circuitry, and work all the time on batteries without having to be booted up — all without requiring a big computer.

## PHIL'S STORY

Limor and I live at the same address, and after we developed the Tweet-a-Watt, she would look at the power graphs it created each morning. She saw these huge spikes every 2–3 hours at night, so she asked me what was going on. I needed to admit that I have a hard time sleeping, and those spikes were when I went to the other room (my office) and turned on my giant 30" monitor. This project now gives away some of my sleep problems — if I could fix those, I'd likely consume less power.

## RESOURCES

Tweet-a-Watt GAE web app: wattcher.appspot.com
Tweet-a-Watt Twitter account: twitter.com/tweetawatt
Limor's original Tweet-a-Watt tutorial:
   ladyada.net/make/tweetawatt

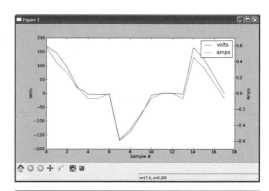

## Tweet-a-Watt Code

I wrote the receiver software in Python, which is a fairly easy-to-use scripting language that has versions for all OSes and tons of tutorials free online (python.org). There's a handy XBee library for Python you can include at the top of your code so you don't have to count bytes and calculate checksums to unpack XBee's packet format.

The code is one big loop that continuously grabs packets from the PC's serial port, then calls the XBee library to retrieve just the sampling data coming into the XBee's analog pins AD0 and AD4. Then it normalizes the data to get real volts and amperes, throws out the first and last readings from each burst, and averages the remaining 17 samples to derive a single numerical voltage reading.

Given the voltage readings, multiplying by amperes (current) gives us watts, and multiplying again by time gives watt-hours. For logging, the samples are averaged again into readings every 5 minutes and stored into the log file *powerdata.csv*, one line per sensor reading, following the format: Year, Month, Day, Time, Sensor Number, and Watts. The values are comma-separated so you can import the log into any spreadsheet program.

For the tweets, the code passes your Twitter username, password, and message strings into the Twitter API by calling the *python-twitter* library (developer.twitter.com).

Our web app is live and running at wattcher. appspot.com.
                                        —*Limor Fried*

# ONE-TON LINEAR SERVO

## Mod an electric jack to do your bidding.
### By Windell H. Oskay

Photography and illustrations by Windell H. Oskay

RC hobby servomotors were just made for hacking. What's not obvious is *how much* you can hack them — with a few tricks, you can use a servo to control almost anything.

Hobby servos (Figures A and B, following page) consist of a DC gear motor, a potentiometer (usually 5K), and a control circuit. The output shaft is the output of the geared motor. The shaft also turns the pot, which returns a position-dependent voltage to the control circuit. When the servo receives a command to move to a new position, it runs its motor in the necessary direction until the pot indicates that the output shaft has reached the desired location.

What if you want motion more powerful or complex than a standard servo can give — can you do it? Certainly! Just replace the little motor with a bigger one (and whatever driver is needed) and/or replace the pot with one that senses the movement that you really want.

Let's get started with extreme servo hacking. At Evil Mad Scientist Labs, we've modified a standard hobby servo to control an automotive jack. The result is a powerful, sub-$100 actuator that can move a heavy load with precise control, with a total stroke of 5"–10". It's a powerful weapon in the maker's arsenal for theatrical and Halloween props, CNC projects, and robotics applications.

## 1. Hack the jack.

We'll start with a 1-ton, 12V automotive scissor jack (Figure C). The controller has buttons labeled "up" and "down." Open the controller to find the backside of the up button. With your continuity tester, find the 2 pins of the button that become connected when you press it (Figure D). Solder 2 long wires to

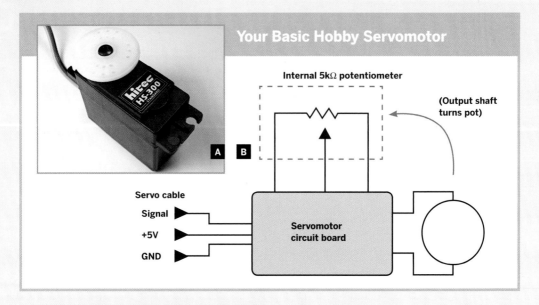

## Your Basic Hobby Servomotor

Internal 5kΩ potentiometer

(Output shaft turns pot)

A B

Servo cable

Signal

+5V

GND

Servomotor circuit board

---

## MATERIALS

1-ton, 12V scissor jack $60 at major auto parts stores
Standard hobby servomotor any type
Servo controller or R/C receiver
5kΩ, 10-turn potentiometer such as Digi-Key
    part #3590S-2-502L-ND, digikey.com
5V relays (2) Hamlin HE3621A0510 (or similar),
    Digi-Key #HE207-ND
2N3904 transistors (2)
4.7kΩ resistors (2)
47µF, 16V capacitors (2)
500Ω 1-turn trimpots (2) such as Digi-Key
    #3362M-501LF-ND
LEDs (2, optional)
75Ω resistors (2, optional)
Insulated copper hookup wire
Protoboard such as BG Micro #ACS1053 or
    RadioShack #276-149
Masking tape
String
¼" ID (internal diameter) washer

Optional, depending on configuration:
Wood for mounting setup
Spring or big rubber band
1lb weight e.g., a can of soup
Screw eyes (2)

## TOOLS

High-current 12V DC power source such as a car
    battery, car cigarette lighter output, or equivalent
Soldering iron and solder
Desoldering braid
Wire cutters and strippers
Multimeter (continuity checker/ohmmeter)
Screwdrivers

---

to those pins. Repeat for the down button, adding 2 more long wires across it (Figure E). Reassemble the controller (Figure F). You may need to make a little notch for the wires to escape.

Now, if you touch together the far ends of the 2 wires across the up button, the jack will believe that you pressed the button itself. This scheme is minimally invasive, and it lets the jack's top and bottom limit switches still function normally.

## 2. Get the servo to serve you.

Next, you need to eviscerate a servo. The best candidate is one with a bad motor, a broken case, stripped gears, or a bad pot. Unscrew and open the case to get at the printed circuit board (PCB). The appearance may vary tremendously, but the features shown in Figure B are universal: an input (signal and power) cable, 2 outputs that go to the motor, and 3 wires that go to the pot (Figure G).

Cut these latter 5 wires in half, carefully noting which 2 went to the motor and which one went to the middle (wiper) terminal of the pot. (Some servos have the motor and/or pot soldered directly to the PCB without wires. In that case, unsolder and remove the part(s) instead of cutting the wires.)

You're replacing the servo's internal single-turn pot with an external 10-turn pot. Solder 3 long wires — with a different color for the wiper — from the original pot's contact tabs to the 10-turn pot, connecting wiper to wiper. (The wiper pin of a 10-turn pot is usually at the end opposite the shaft.)

To control the jack, you'll use little relays driven by

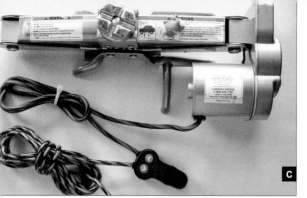

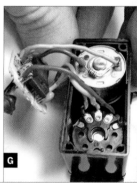

Fig. C: An ordinary 1-ton, 12V scissor jack.
Fig. D: Find the "up" and "down" button pins.
Fig. E: Solder 2 pairs of wires to the button pins.

Fig. F: Reassemble the controller.
Fig. G: Connect 2 wires to the servo's motor, and 3 to its pot. Fig. H: Add 2 relays to control the jack.
Fig. I: Construct 2 output drivers.

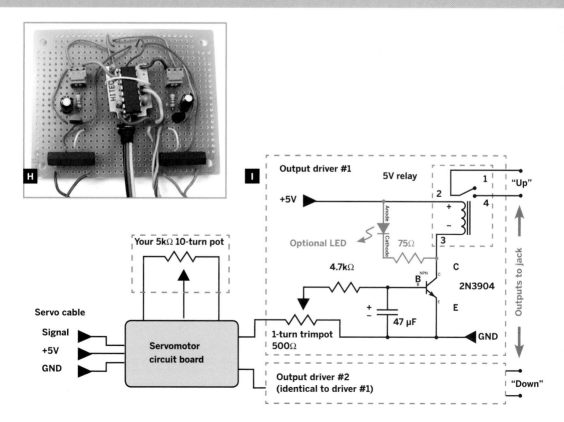

**Your 5kΩ 10-turn pot**

Servo cable

Signal

+5V

GND

**Servomotor circuit board**

1-turn trimpot
500Ω

**Output driver #1**

+5V

5V relay

Optional LED

Anode

Cathode

75Ω

4.7kΩ

47 µF

1    "Up"

2

4

3

NPN

B

C

E

C

E

2N3904

GND

Outputs to jack

**Output driver #2
(identical to driver #1)**

"Down"

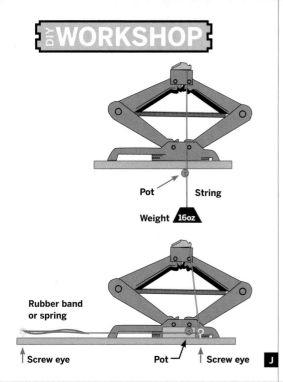

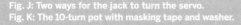

Pot     String

Weight **16oz**

Rubber band
or spring

Screw eye     Pot     Screw eye

**J**

**K**

**L**

Fig. J: Two ways for the jack to turn the servo.
Fig. K: The 10-turn pot with masking tape and washer.

Fig. L: Center the trimpots, center the jack and the
pot, then tension the string.

adjustable, single-transistor driver circuits (Figure H).
 Begin by identifying which of the 3 wires on the servo's control cable are power (+5V) and ground, and solder connections to those points.

Construct output driver #1 as shown in Figure I, watching the orientation of the capacitor (its negative side is marked), the transistor (its legs read E-B-C when you can read the writing), and the relay (its pin 1 is marked). Take either wire that previously went to the motor as the driver's input, and connect the driver's output to the 2 wires across the up button on the jack. An optional LED can be added for debugging.

Make an identical output driver #2 with the other motor wire, and connect its output across the down button (Figure H).

## 3. Finish the feedback.

The jack's movement must turn the pot in order for servo control to take place. Two working geometries are sketched in Figure J. You can tie a string to the top of the jack, wrap the string one turn around the shaft of the pot, and then hang a weight below to maintain tension. Or, a more general scheme is to wrap the string through a screw eye, around the pot, and then attach the other end through a rubber

band (or spring) to a second screw eye.

For either method, the pot's shaft needs some friction: wrap regular masking tape around the shaft 10 times, and then slide a ¼"-ID washer over the tape (Figure K). This holds the tape in place and stops the string from rolling off the edge of the tape.

## 4. Get set, go!

Center both 500Ω trimpots in their ranges, and hook up your 12V source for the jack. Manually adjust the jack to middle height, move the 10-turn pot to mid-range, and then tension the string (Figure L).

Turn on your servo control signal and see if it works. The jack's movement should "lock up" wherever you want it to stop. If it doesn't lock up — the jack moves to one end of its travel regardless of signal — then the feedback is backward. Fix it by swapping the 2 non-wiper wires going to your pot.

The 2 trimpots adjust the "dead band" of the servo; tune it between a loose lock, where precision is lower, and a too-tight lock, which may oscillate.

As a child, Windell H. Oskay was inspired to pursue a Ph.D. by *Doctor Who* reruns. Now he helps run a website called Evil Mad Scientist Laboratories (evilmadscientist.com).

# $30 MICRO FORGE

## Make your own nails and other small iron parts. By Len Cullum

Photography by Len Cullum

I am a woodworker by trade and spend my days building Japanese-style architectural elements and structures. Because of my chosen niche, I occasionally need a piece of hardware that's impossible to find in this country. Sometimes I ask friends in Japan to track it down. But other times, if the piece isn't too complicated, I'll make it myself.

Recently, I needed to make 500 old-style Japanese nails. In the past I would simply fire up a propane plumber's torch, hold each piece over it until it was glowing, and then hammer away. That was fine when a dozen nails were all I needed, but this time I knew I needed a more efficient way to heat all those nails, so I built a micro forge.

### 1. Drill the brick.

Firebricks are really soft and easy to cut. So soft, in fact, that you can dig into them with a fingernail, but they are still rated to 2,300°F. Ordinary wood/metal drill bits go through them like butter.

The nails I was making were 2½" long, and I wanted to keep the forge chamber as compact as possible, so I used the brick's biggest face (4½"×9") as the front, which allows for the shallowest chamber. Make a mark on the front face, 2¼" in from one end, and centered 2¼" from either side (Figure A, following page).

Wear a dust mask. Set the drill press to its lowest speed (mine was 250rpm) and drill a hole 2⅛" wide and 2" deep, centered on your mark. Drill slowly — you don't want this stuff flying all over the place (Figure B). With a slow speed, most of the dust should stay in the hole until you dump it out.

Locate and drill the ¾" flame hole, which comes in from the side. I chose to locate it at the top of the forge chamber (Figure C) so that the curved sides

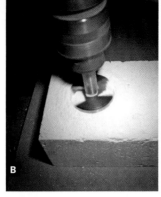
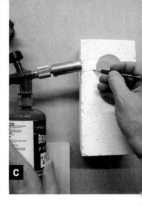

A

B

C

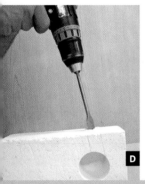

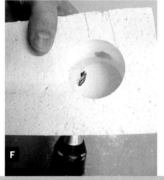

D

E

F

G

Fig. A: Mark the chamber location on the firebrick. Fig. B: Drill the firebrick slowly. Fig. C: Position the flame hole so that flame will circulate around the chamber. Fig. D: Drill the flame hole.

Fig. E: Flame hole and chamber drilled. Fig. F: Drill the vent hole in back. Fig. G: To build the frame, first mark the edge of the aluminum angle on the plate.

## MATERIALS

*The firebrick is a specialty item; the rest is available at your local hardware store.*

**K-23 3" insulating firebrick (soft), 3"×4½"×9"** seattlepotterysupply.com **item #31414**
**7/16" threaded rod, 36" lengths (2) You'll cut them in half.**
**7/16" nuts (16) These should fit the 7/16" rod but have a smaller outer diameter than the 3/8" washers.**
**3/8" washers (16)**
**1/16"×1" aluminum angle, 24" length**
**1/16" aluminum plates, 4"×6½" (2) Cut to this size.**
**16d finish nails (3)**

## TOOLS

**Drill press or hand drill**
**Hacksaw**
**Drill bits: 2⅛" Forstner or spade bit, ¾", ½", ¼"**
**Wrenches (2) for tightening the nuts**
**Propane plumber's torch preferably with a jumbo tip or other high-volume flame**
**Dust mask**

would cause the flame to circulate, hopefully heating everything inside more evenly. I tried to drill this hole to match the angle of the torch tip (Figures D and E).

The last step is to drill the small vent hole in the back (Figure F). I made this hole at the bottom of the chamber at an angle in order to coax more convection from the flame.

## 2. Build the frame.

My first micro forge attempt was just the firebrick with the holes in it. This worked great for about a minute, then the heat caused the brick to crack and start to open like a clam. It needed a frame.

Because I wanted to make it fast and simple (and my welding skills are a bit rusty) I decided to use off-the-shelf parts that could be easily cut.

Start by cutting the 1" aluminum angle into four 4" lengths. Align a piece on the 4" edge of the aluminum plate and stand the brick against it.

Place a second angle piece on the other side of the brick and mark the plate along the angle's edge (Figure G). Then clamp the 2 plates together and cut them to size along the line (Figure H).

Cut the threaded rods in half and file or grind the cut ends to remove any burrs.

Now you'll need to lay out the corner holes that

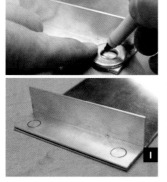

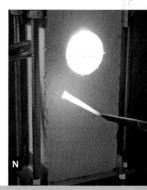

Fig. H: Cut the top and bottom plates together. Fig. I: Trace washer holes on the ends of aluminum angle pieces. Fig. J: Drill the angle and plate pieces clamped together. Fig. K: Bottom plate and angle supported and secured by nuts on threaded rods. Fig. L: Firebrick in place between the top and bottom plates. Fig. M: Grate made from cut nails tapped down into the brick. Fig. N: Micro forge in operation.

will accommodate the 4 threaded rods. To assure a proper fit, place a washer in each corner and trace the hole (Figure I). Then clamp the angle to the plate and drill a ½" hole in each corner (Figure J).

## 3. Assemble the forge.

At this point you need to find the height of your torch tip. Measure from the bottom of the tip to the bottom of the tank, then subtract the height of the flame hole. The remainder is your leg length.

Thread a nut onto each of the 4 rods to the leg length, then add a washer. Slip the short end of the rods through the angles and plate, then add another washer and nut (Figure K). Don't wrench-tighten yet because you'll need to adjust everything for height and make sure it's level once it's all assembled.

Stand it up and add the brick. Adjust the angles so that they fit snugly against the brick without crushing it. Stand the torch next to it and make sure the flame hole and tip are lined up.

When everything is sitting right, spin 4 more nuts and then washers on until they're just below the top of the brick. Add the other 2 angles and the top plate, with washers and nuts (Figure L). Carefully hand-tighten the upper hardware so that every-thing is holding the brick without putting too much

pressure anywhere. Wrench-tighten everything.

Finally, add the grate that the parts rest on while heating. To keep it simple, I used three 16d finish nails. I cut their heads and points off, to a length of about 2¹⁄₁₆", put them in the forge chamber, and gave them light downward taps to seat them in place (Figure M). This made a grate low enough to allow me room for moving things in and out, and wide enough to hold 4 nails at a time.

## 4. Fire in the hole.

You're ready to fire it up! Place your micro forge and torch on a flat surface separate from the one you'll be pounding on, light the fire, and watch it make your metal glow (Figure N).

As of this writing, I've made more than 1,000 nails in this little forge and it's still going strong. The only change I might make would be to add all-thread connector nuts as feet. I've discovered that propane tanks vary in height a bit, and the addition of feet would make it easier to adjust the height of the forge.

Len Cullum (shokunin-do.com) is a woodworker in the Japanese style in Seattle. He makes shoji doors and windows, garden structures, and architectural elements.

# DIY OUTDOORS

# BARREL WATER COLLECTOR

## Make wine into water (sort of).
### By Chris Barnes and Michri Barnes

Many people let the rain that falls on their roof run away, then they use drinking water piped in from afar for washing floors and watering plants. Here's a handy, mosquito-proof rain barrel we put together that buffers 55 gallons of water and adds a handsome accent to our yard. It's especially valuable during droughts, and if you're in a rural area with wells and electric pumps, it also means being able to flush the toilet when the power goes out. The barrel and fittings are also suitable for potable water, but don't store graywater, or pathogens can grow.

Our barrel sits under an eave of our house, where even on foggy days it collects water that trickles down. You can also put it under a downspout, or anyplace else outside where it will capture water.

And if you're really ambitious, you could have a series of barrels and move the pump from one to the next, or even interconnect them.

## 1. Make holes in the barrel lid.

Lay out the following holes on the barrel's cover and drill them with the hole saw. You need one hole near the edge for your pump's down tube, and two more for collected water to drain through.

Use a strong drill, and draw the hole saw out to clear away sawdust every once in a while. It also helps if you remove some wood from the hole by chiseling across the grain at the edges (Figure A, opposite page).

When you smell wine, you're almost there. One of our plugs fell in, but that's no disaster; it just means there's some wine barrel in our wine barrel.

Cover the drain holes with screen to keep out mosquitoes and debris. I cut two 4" squares with a straightedge and utility knife, then folded the edges in and stapled them down (Figure B).

Photography by Michri

## MATERIALS

*Get the barrel and hand pump first, then determine the sizes of other parts to fit.*

**Barrel, covered** We got ours for $20 from a local winery, but you can also use a whiskey or pickle barrel. Or check garden supply, home improvement, or grocery stores. Barrels get leaky and eventually fall apart when they dry out, so if you don't get to your project right away, put some water in the barrel or hose it down occasionally.

**Hand pump, aka pitcher pump** $40 from Northern Tool + Equipment (item #108980, northerntool. com) or check your local hardware store

**Bung or expansion plug** from the winery (or other barrel source) or a hardware store

**PVC pipe, about 3' long** Its diameter must fit the pitcher pump; ours was 1¼".

**PVC foot valve** to fit pipe

**PVC adapters** to fit pipe:
   Male pipe thread (MPT) to slip
   Female pipe thread (FPT) to slip

**Teflon plumbing tape and TFE (teflon) pipe thread sealer paste** You can use just one or the other, but my professional plumber friend uses both.

**PVC purple primer**

**PVC cement** such as Christy's Red Hot Blue Glue or Gorilla PVC

**Wire screen, bronze or stainless steel, about 5"×9"** Check the scrap bin at your hardware store.

**Screws** to attach the pump to the barrel

**Spray-can insulating foam** We call it funny foam.

## TOOLS

**Drill**
**2½" hole saw**
**⅛" drill bit**
**Mallet or hammer**
**Chisel**
**Staple gun and staples** or small tacks
**Utility knife**
**Straightedge**
**Hacksaw or PVC saw** A PVC saw cuts straighter.

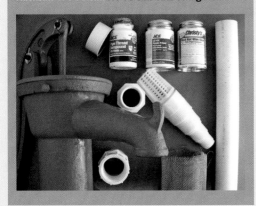

Fig. A: It helps if you remove some wood by chiseling the hole saw cuts; be sure to cut into the grain.
Fig. B: The uncovered hole in the top accommodates the pump, while 2 screened holes let water drain in.

## 2. Install the pump.

Our hand pump came with a check valve, but its quality was questionable, so we installed a foot valve to keep the pump primed. (It sucks when all you want is a bucket of water, and you need a bucket of water to get it.) The valve was multi-size, so we first had to cut a section off the end so it would fit our 1¼" pipe.

First, screw the pipe adapters onto the hand pump and the foot valve. Wind teflon tape 3–5 times around the threads of both adapters, in the direction you'll be screwing. Apply TFE paste on top of the tape. Then screw the foot valve into the FPT-to-slip adapter, and screw the MPT-to-slip adapter onto the pump. Screw both as tight as you can with your hands.

If you find that you lose the prime on your pump, check and tighten these connections, but be careful; overtightening PVC fittings can cause cracks.

In a well-ventilated area, liberally apply PVC primer and then PVC glue to both the PVC pipe and the slip fitting on the foot valve. Attach the two by giving the pipe a slow half twist and a quarter twist back as you push it in.

Set the pump down and measure how high its slip fitting hangs. Set the 3' pipe into its hole in the top of the barrel, mark the barrel's height, and add the slip fitting distance (Figure C). Cut the pipe to that length, and deburr the edges with a knife. Glue the pipe into the pump's slip fitting.

Set the pump in place and securely screw it down (Figure D).

## 3. Seal and vent.

Seal the bunghole on the barrel's side. Tap a wooden bung in with a mallet, or push an expansion plug in as far as you can, and tighten the wing nut. If the bung leaks at first, it will probably swell up and seal.

Spray insulating foam to fill the gap around the down tube. You may want to wait until you've used the barrel for a couple weeks and know that everything is working properly. Don't touch the foam while it's still wet; it makes a mess. After it dries, cut off the excess with a knife.

**TIP: Once you've started using the can of foam, you have to use it all, so if you have any other holes to fill, like around your doorjamb, do it now. Or you can make a giant fake dog doo, a funny hat, or a combination of the two.**

Finally, drill 8 or more ⅛" vent holes around the barrel, about 3" down from the top. These let air escape from the barrel during a heavy downpour, and prevent standing water on top when the barrel's full.

That's it! Once we finished, we couldn't wait for the rain to test the pump, so we put some water in the barrel with the garden hose.

We've been using water from the barrel ever since for our plants, pets, chickens, and ducks. It's also been nice to have around for flushing toilets when the pipes freeze.

Chris Barnes (cpaynebarnes@gmail.com) and Michri Barnes manufacture sustainably harvested wood knitting needles on the North Coast of California. They garden, keep birds and bees, build their house, and love to play outdoors.

Fig. C: Measure the down tube to extend from the adapter under the pump valve to the barrel's bottom.
Fig. D: Wear proper attire while screwing the pump onto the barrel top; a MAKE shirt always works!

# RAINWATER TOILET FLUSH

## Save your tap water and let the rain in.
By Eric Muhs

Photograph by Eric Muhs

I found several nice, free 55-gallon steel drums a few years ago. I used some of them as musical drums in Metal Men, a noisy, homemade instrument my friend John Hawkley and I made together. I gave another to students to make into a smoke ring generator.

The last two I made into a rain barrel system for collecting the runoff from my roof. Eventually I figured out the best use for them — flushing!

### Catching the Rain
For my rain barrel system I removed part of a downspout, and replaced it with an accordion-folded plastic extension from the hardware store. This runs into a funnel, and into a 3" hole conveniently provided in the barrel lid.

Overflow from the primary barrel runs via a hose outlet at the top of the barrel to a second barrel.

**MATERIALS**

**55gal steel drums (1 or more)**
**Rain gutter downspout extension**
**Funnel**
**Cinder blocks or other support**
**½" hose bib**
**Hose**
**2-way hose divider and adapters as needed**
**Plumber's goop**
**Plastic gaskets cut from the lid of a 5gal bucket**

Both barrels have a lockable lid, an important feature to keep the water relatively clean and safe from drownings by squirrels or small children. In certain areas of the country, mosquitos would be another reason to keep the lids closed.

Linked in this way, the barrels gather water pretty effectively: they fill easily after a rainy day,

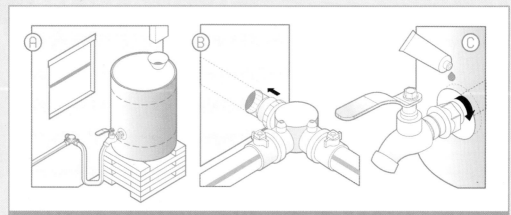

Fig. A: Barrel under the downspout has an on/off valve to feed the hose, which runs inside through the hole in the wall. Fig. B: The hose divider under the toilet lets you supply it with either rainwater or municipal water.

Fig. C: Plumber's goop between the plastic gaskets and the curved metal barrel makes the valve seal drip-proof.

with very little gutter debris making it in. Originally, I thought they'd work great for my garden.

Here's the problem: this system was almost useless for irrigation. When it rains in Seattle (October to May, every day), there's no need for rain barrels. When it doesn't rain, the rain barrels are emptied in one watering of the garden. And the watering itself is slow and hard to distribute, because of the extremely low pressure of the gravity-fed water from the barrels.

Could I repurpose these barrels? Yes! I reorganized them into a toilet-flushing system. This would allow me to use the harvested water throughout the year.

## Rain Barrel to Toilet Tank

I built a platform on top of a stack of cinder blocks, under the downspout closest to the bathroom (Figure A). A hose from the lower outlet of the barrel ran through a hole in the wall near the toilet, and into a two-way hose divider below the toilet (Figure B).

I can flip the valves to fill the toilet with city water, or with harvested roof water. Water siphons easily into the toilet tank from the barrel, and the regular float valve inside the tank cuts off the barrel water just as it did before.

## Sealing the Barrel Outlet

Attaching hose valves to the metal barrel was the hardest, costliest part of this project. There are kits available, but I chose to go with off-the-shelf components. Drilling a hole large enough for a ½"

hose bib was noisy and tedious with the biggest drill bit I could fit into a hand drill. Standard pipe and hose hardware don't sit well on the curved surface of a barrel, and it took several tries and lots of dripping before I came up with a system that works reasonably well.

I cut 2 large plastic gaskets from the lid of a 5gal bucket, and fit them onto the hose bib as I was pushing it through the hole in the barrel, so the gaskets sat against the barrel, on the inside and outside.

While these were still loose, I filled the space between each gasket and the barrel with plumber's goop, then turned the hose bib many times to smear the goop and get good coverage all the way around the fitting (Figure C). When I tightened the fitting down, and let the goop cure overnight, this provided a good, drip-proof seal.

## Going Bigger

I found that my family of four could easily drain a full barrel dry over the course of a stay-at-home weekend. Between rains, we were running dry regularly. So I recently built a second raised platform, and doubled my storage capacity. The 2 barrels are connected by a short hose at their bases, so they have the same fill level.

---

Eric Muhs is a physics teacher, inventor, photographer, musician, father, and adventurer. He takes aerial pictures with kites, builds musical instruments, and travels everywhere.

Illustration by Julian Honoré

# SOLAR HYBRID HOT TUB

 A free and easy thermal assist
from the sun. By Eric Muhs

Photography by Eric Muhs

Back when I lived in California, I heated the water for my house and hot tub using some 1970s "energy crunch"-era solar thermal collectors that I found at a junkyard. They were a nice design from Israel, with a large plexiglass collection surface and an insulated horizontal tank with its outlet up top, where the hottest water rises.

Those restored units were all the water heating we needed for about 9 months of the year, and I installed them on a hillside rather than the roof for easier maintenance. (Alternative energy means lots of repairs; if you're harvesting energy from the environment, your equipment will be, well, out in the environment.)

Then I moved to a newer house in Seattle with a fancy automatic hot tub on the rooftop deck. But I was appalled when I saw that it cost up to $40 in electricity each month to heat this mass of water

and leave it covered outside overnight.

So I came up with this simple and inexpensive solar system that adds heat to the hot tub during the day so that the main electric heater doesn't have to work as hard after the sun goes down.

## Solar Collectors

The system heats the water by running it through coils of black vinyl hose inside 2 solar collector boxes. A small solar-powered pump draws water out of the tub, runs it through the coils, then dumps it back in. The tub's original heater and thermostat are not altered, but the heater switches on less.

For the solar collectors, I built 2 open 3'×3' boxes out of plywood reinforced with blocks of scrap wood in the corners (I had just 1 box originally, but expanded the system later). I put the boxes together using 1¼" wood screws and glue, then caulked the

## MATERIALS

½" plywood sheets, 8'×4' (2)
Clear plexiglass sheets, 3'×3' (2)
½" black vinyl hose, 100' rolls (3)
Scrap wood **I used some 2×2 lumber ends.**
Matte black paint
Small pump **I used a 12V DC marine pump; a $20
    fountain pump that runs off AC would also work.
    The less water the pump moves, the hotter it
    heats, so no need for a monster.**
Solar panel **to run the pump; I got mine from eBay.**
Hose fittings:
    Female ¾" garden hose to male ½" hose barb (2)
        to connect the pump to the vinyl hose. May
        vary depending on your pump.
    Male to male ½" hose barb (3+) to join vinyl hose,
        plus extras for repairs
Wood screws, 1¼" or so (24)
Wood screws, ¾" or so (48) with washers (48)
Carpenter's glue
Caulk

## TOOLS

Saw, drill and drill bits, caulking gun, paintbrush,
screwdriver, utility knife

Fig. A: Collector box with black hose coiled inside.
Fig. B: Intake and outlet hoses dangle in the tub.
Fig. C: Solar-powered pump sits under the tub.

cracks and painted them matte black inside and out. For the vinyl hose inlet and outlet, I drilled a hole in the side of each box near the bottom (Figure A).

I tucked the pump under one side of the hot tub, and used hose barb fittings to connect each end to black vinyl hose. Then I connected the rest of the hose together, coiled it inside both boxes, and dropped each end into the tub, separated slightly (Figure B). The pump sits on the inlet side of the hose, but I don't think it matters.

I fitted plexiglass tops onto the boxes with ¾" screws, washers, and more caulk, and tilted the boxes up on the deck across from the hot tub, to lean south for more sunlight.

The vinyl hose is inexpensive and easy to repair in case of freezes: just cut out the broken part and insert a barbed connector.

**TIP: Dipping the hose into very hot water for 15 seconds makes it easier to slip a fitting inside.**

## Pump Power

My original system had just 1 collector and used an aquarium pump on a 1-hour timer. But the pump ran even on cloudy days, when it drew hot water out of the tub to get cooled — which was not the idea.

I thought of adding a light-sensitive switch to prevent this, but instead chose to power a new DC pump with a solar panel. The pump sits under the tub (Figure C) and only runs in bright sunlight, when the boxes heat the water most efficiently.

## Results

This system doesn't provide the same level of heating as my old setup in California. Its surface area is smaller, and with less sun and colder temperatures, the Seattle weather doesn't help. As my old California neighbor Steve said, "Here, you can just throw hose on the ground."

But on sunny days, the days when your parked car is hotter than the outside air when you get in, it works great. The water returns to the tub 2° or 3° hotter than when it left, which may not sound like much, but it adds up.

The entire system cost me about $70 for parts and materials, and it will keep paying back as fuel prices rise. In warmer, sunnier climes, systems like this can actually overheat your tub. Unlikely in Seattle, but I did get my hot tub too hot once or twice in California. It cools off pretty quickly with the top off. As my physicist friend Heyward once said, "Heating water is not rocket science."

# BIKE REPAIR STAND

## A cycle support from the plumbing aisle.
By Shaun Wilson

I looked around town for a bicycle repair stand, and the cheapest one was $150. Yikes! Rather than resign myself to flipping the bike upside down on its seat and handlebars, straining my back, and always having to work upside down, I made my own repair stand out of galvanized pipe for about $30. It's easy to take apart and reconfigure with new pieces, so I've refined the design through multiple iterations. Here's the latest and most stable version of the hardware.

You can find almost all the materials in the plumbing section of a hardware or home improvement store, or if you're super resourceful, you may be able to scavenge it all for free.

### 1. Build the base.
Begin by screwing together the legs, which will be mirror images of each other, around 2 tee fittings.

**MATERIALS**

½" galvanized pipe, double-threaded, in the following lengths: 48", 18" (2), 10" (5), 8" (2), 3" nipple (2)
½" galvanized pipe fittings: tees (3), 90° elbows (3), 45° elbows (2), end caps (3)
Hose clamps (2) to fit ¾" to 1½" hose
1×2 pine board, 18" long
Screw-in utility hooks (2) the heavy-duty kind for hanging things up in the garage
Pipe hangers (2) I used TouchDown pipe clamps. These are plastic, with a bridge that ratchets down over the pipe like a zip tie and a metal screw that secures it to the wood.

**TOOLS**

Flathead screwdriver
Drill and drill bits

DIY **HOME**

Fig. A: Base of the stand. Galvanized pipe can be a heavy-duty Erector set. Fig. B: Cross-members are hose-clamped to the 48" upright to prevent it from leaning. Fig. C: Cross-members are connected to the base and upright with hose clamps. Fig. D: Pipe hangers connect the horizontal pipe to the wooden board. Utility hooks screwed into the board hang down to hold the bicycle.

The stem of each tee points up and connects to a 3" nipple, which in turn connects to a 45° elbow. Horizontally, the parts run: end cap, 10" pipe, tee, 8" pipe, 90° elbow. These can now be used as sweet weapons. No, put them down.

To complete the base, connect each 90° elbow to another 10" pipe and join them in the middle with another tee (Figure A). These connections don't need to be extremely tight, so just hand-tighten them. I chose not to buy a pipe wrench, figuring that the geometry of the stand itself would keep the pieces tight.

## 2. Add the upright and supports.

Now, up we go! Screw the big 48" length of pipe into the stem port of the middle tee, so that it sticks straight up. Screw the 18" pipes into the 45° elbows on the base, so that their free ends just meet the upright pipe, and attach them to the upright with 2 hose clamps (Figure B). These serve as cross-members, to stop the upright from rotating or leaning (Figure C).

Screw a 90° elbow to the top of the upright, attach a 10" pipe, and finish with an end cap to prevent small creatures from getting inside and building tiny, tiny cities. That's it for the pipe, so

tighten everything up and restore the symmetry.

If you decide you don't want a repair stand, this now functions as a hat rack. For one hat.

## 3. Install the bike hooks.

To hold the bike, I screwed 2 strong hooks into the ends of an 18" length of 1×2 pine (drill pilot holes first to prevent splitting). Then I screwed 2 plastic pipe hangers onto the middle of the board and clamped them around the end of the 10" pipe, with the hooks hanging down (Figure D).

The bike can now be lifted and set into the hooks. This is where you'll find out if your connections are tight. Note that you don't have to find the center of gravity for the bike to stay level, as you would if you used a single clamp in the center. If your bike has cables running under the top tube, you'll have to thread the hooks between the tube and cables.

Shaun Wilson is a project manager/software engineer in Anchorage, Alaska, whose lifelong interest in building things is most likely related to watching *This Old House* on PBS. He is currently replicating some of the projects from *Eccentric Cubicle* (Make: Books).

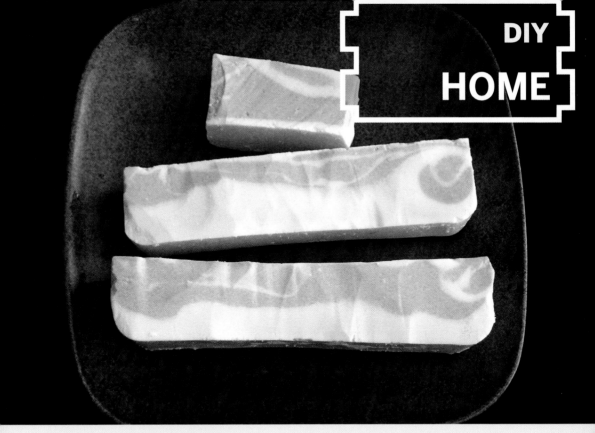

# HOGWASH

 ## How to make bacon soap, from actual bacon! By Tim King

I wanted to see if it was possible to make soap from bacon fat. The bonus challenge: to make the soap look like bacon.

## Prepare the Bacon Soap

**1.** Melt the bacon fat on the stove (Figure A, following page). Go slowly and don't allow it to boil or sizzle. Skim off any particles or debris that float to the surface of the melted bacon fat.

**2.** Filter the melted bacon fat by pouring it through cloth or paper towels into a large, clean metal can (Figure B).

**3.** Still warm, your bacon fat should now look uniform, clear, and junk-free. Measure your bacon fat, then calculate the proper amount of lye — 1 part lye to 7½ parts bacon fat.

**MATERIALS AND TOOLS**

Bacon fat
100% lye (sodium hydroxide)
Purified water and ice
Liquid smoke flavoring (optional) **for fragrance**
Red food coloring (optional)
Stove
Pyrex baking pan
Chemical/solvent/heat-resistant plastic bowl
Measuring cup
Cotton cloth or paper towels
Wooden or stainless steel spoon
Metal can
Kitchen thermometer

*Use the following volume ratio:*

   7½ parts bacon fat
   1 part crystalline lye
   2 parts water

Photograph by Tim King

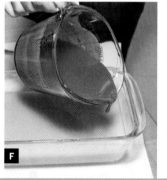

**DIY HOME**

Fig. A: Melt the bacon fat (Mmm!) over a stove.
Fig. B: Filter the melted fat, to ensure Ivory Soap purity.
Fig. C: Measure lye crystals (wear eye protection).
Fig. D: Pour warm lye mixture into the warm bacon fat.

Fig. E: Stir the mixture until you see trace (white streaks) on the surface. Fig. F: Add red dye (optional) to a small portion, to honor the original bacon.
Fig. G: Cut the partially cured soap into desired shapes.

**4.** Measure 2 parts purified water. Feel free to include some ice in it, as the lye will get very hot. In a heat-resistant, nonmetal container, pour in your 2 parts water, then slowly pour the crystalline lye into the water.

⚠️ **WARNING: Lye fumes are noxious and can cause serious injury. Make sure to wear rubber gloves and eye protection when handling lye, and to mix it in a well-ventilated area. Also, the lye and water mixture will get very hot, so be sure to mix them in a proper workspace, and in a heat-resistant nonmetal container.**

**5.** Take a break and wait for the lye-water mixture to cool down. Do not leave the lye mixture unattended. When the lye-water mixture is around 100°F (the temperature of a hot shower), check your bacon fat and see if it's around the same temperature (if not, heat it on the stove until it is). Then slowly pour the warm lye mixture into the warm bacon fat (Figure D).

**6.** Stir the mixture for up to 2 hours as it cools (Figure E). When the mixture begins to show "trace" (white soap stripes on the top surface), move on to the next step.

**NOTE: Mixing can take several hours. If your mixture is still liquid after 2 hours, move on to the next step and hope for the best.**

**7.** While the proto-soap mixture is still a thick liquid, but showing signs of trace, you can mix in liquid smoke flavoring as a fragrance to enhance the bacon smell. Also, if you want to make your bacon soap look like bacon, now's the time to pour a small portion of the mixture into a separate container and add a bit of red dye.

Pour the beige, thick proto-soap into a pyrex glass pan, then take the red-dyed proto-soap and pour stripes into it to make it look like bacon (Figure F).

**8.** Let the soap set or "cure" for at least 36 hours, then cut blocks into desired shapes from the greater slab (Figure G). Do this while it's still relatively soft. Allow the soap to continue to cure for at least 2 weeks.

**9.** Enjoy your soap. Everybody loves bacon (the meat of the gods), and now we can bathe with it.

Tim King (tim.king.phd@gmail.com) is a professor of anthropology and archaeology who loves re-creating ancient technologies. He's also a big fan of bacon.

Photography by Ed Troxell

# 1+2+3 The Two-Person Shovel
By Josie Moores

For anyone who's spent the better part of a day shoveling, you know how utterly exhausting it can be. Here's a way to get those big digging projects done without a single backache or blister: witness the two-person shovel. All you need is a friend and a length of rope.

YOU WILL NEED

Ordinary shovel
Sturdy piece of rope
Stick or similar handle (optional)

## 1. Cut the "second" handle.
Cut a piece of rope that's about the length of your leg.

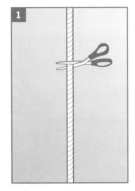

## 2. Attach the rope to the shovel head.
Attach one end of the rope near the head of the shovel.

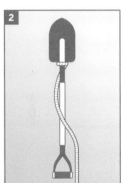

## 3. Make a handle at the top of the rope.
Make a loop at the other end of the rope, and a second loop a few inches away from the end loop. Secure the loops by knotting. Or, if you'd rather have a handle, wrap the rope around the center of a stick.

That's it for the construction part of the project.

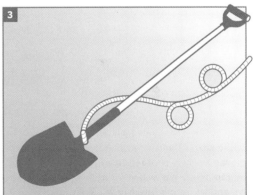

## Use It.
Using the two-person shovel is a bit more complicated than making it, but only because you expect it to be harder than it is. When my husband and I first used it, we were each riddled with guilt that the other must be doing all the work. It seemed too effortless. Only when we switched positions did we believe that neither person was really exerting himself or herself. It's great!

Here's how it works. Person A holds the shovel, and Person B holds the rope handle. Person A places the shovel head on the ground against whatever it is that has to be shoveled (sand, dirt, whatever). He does *not* push, he simply guides the tool. Person B pulls the handle (and thus the shovel head) toward her, scooping up the material to be removed. Person B then lifts the handle upward, and Person A guides the shovel, using the shaft as lever, to where the material needs to be dumped.

Rhythm is the key, and after a few tries, the team can move a *lot* of material without straining.

➕ More info: velacreations.com/2manshovel.html

Illustrations by Julian Honoré; photograph by Josie Moores

Josie Moores lives with her husband and baby boy on an off-grid, fully sustainable homestead. Read about it at velacreations.com.

# LAY OF THE LAND

## Mapping your lot is the first step in designing a homestead.

### By Terrie Miller

Have you ever thought you might like to:

» Grow, raise, and preserve your own food?

» Make observations and do amateur investigations in earth sciences and biology?

» Make structures using natural building techniques like cob or straw bale?

» Build systems for harvesting and caching the water and energy that already flow through the place you live?

If the idea of combining these and related activities to create a sustainable way of life sounds appealing, you may be interested in the practice known as "permaculture."

# WHAT IS PERMACULTURE?

The modern practice of permaculture was founded by Bill Mollison and David Holmgren in the 1970s as the practice of designing and building sustainable human settlements. The permanence of permaculture lies in its sustainability; it's a design system that's based on close observation of nature to create a way of life that has the same resiliency as a balanced ecosystem.

Permaculturists, or "permies" for short, are infused with optimism, a fascination with the workings of living systems and the land, and a delight in experimentation and making things.

Permaculture is a generalist's delight, a practice that encourages dabbling in many of the arts and sciences, and a culture that embraces DIY activities of all kinds: lush gardens, biodynamic farming, herbal medicine, wildcrafting, tracking, and using earthmoving equipment to build ponds and swales, to name a few.

The first step in creating a permaculture design for a piece of property is to get to know the land in detail, ideally at different times of day (and night), through all four seasons, and in a variety of conditions. Creating a hand-drawn map of your property is one way to get to know your land, and your map will be a useful tool for developing a plan of action.

## SKETCH YOUR LOT

To get started, take a walk around your property and notice the shapes of large features such as buildings and fences.

Look for a couple of good anchor points — permanent structures with well-defined edges. You should be able to pull the tape measure straight between these 2 anchor points, and to several other points in your yard.

Draw a quick sketch of the property, noting the features you want to include on your map. Just rough in the shape of your house and other buildings, and don't worry about trying to make it to scale.

Draw your sketch as large as your paper allows, because you'll be adding lots of lines and measurements that will crowd the page.

Measure the distance between your anchor points and write it down on the map. Keep the tape measure anchored at the first point, and measure to as many other points or corners as you can reach, recording them on your sketch. Then move to the second anchor point, and do the same.

Be sure to measure each point twice — once from the first anchor point, and once from the second anchor point — because you'll be using triangulation when you draw your map.

I also like to measure along any edges that are square, like the sides of a house or building (Figure A), because you don't need to triangulate to each corner of your house if it's really a right-angled box.

### TOOLS AND MATERIALS

**Large sheet of paper** in a sketchbook or on a clipboard, to carry around your yard sketching out the property and recording measurements
**Pencils**
**Tape measure** the longer the better. You can get a good 300' reel tape measure for about $30. Most of these are marked with feet and inches on one side, and feet and tenths of a foot on the other side.
**Marking flags and string (optional)** helpful if some of your measurements will be longer than your tape measure
**Architect's scale** Available in most art supply stores with the drafting supplies, they look like 3-sided rulers. Architect's scales are usually better than engineer's scales for home-scale mapping.
**Straightedge or ruler** for drawing lines
**Large, good quality paper** for your final map
**Drawing compass** If you're buying one, get one with an extender for drawing large arcs.
**Tracing paper or vellum**
**Masking or drafting tape (optional)** if you want to do overlays on your finished map
**Pens, colored pencils, markers, and/or watercolor pencils or paints** for coloring and finishing your map

**NOTE: Make sure to pull the tape measure taut over long distances — if it sags into a big curve, the distance you record could be more than a foot longer than it really is.**

You'll probably find that you can't reach every point on your property from these 2 anchor points. That's OK; just pick 2 other points you've measured, and use them as new anchor points. You can have as many new anchor points as you need.

If you need to measure a distance that's longer than your tape (and you can't use new anchor points to work around it), tie a string taut between the points to mark a straight line. Measure as far as you can along the string, and plant a stake flag in the ground, then measure the next section along the string. Repeat until you can add up your segments to get the full distance.

Your sketch will have several measuring lines radiating out from each of the anchor points, and their measurements. It may get quite messy, but don't worry too much about that. If you need to record too many measurements in a tight space, you can always make a larger sketch for that detail area (Figure B).

## MEET THE ARCHITECT'S SCALE

When you have all your measurements written down, it's time to start working on your actual map. First, get familiar with your architect's scale.

Each edge of the architect's scale is marked for a different scale of measurement. For example, on the 3/32 scale, each tick mark of 3/32" represents 1 foot on the map. In this way, the scale saves you from doing the math to convert all your measurements.

If your measurement is 11', you'll use the distance between 0 and 11 along the edge of the scale (Figure C). Note that the scale doesn't start at zero; the marks to the left of the zero are for fractions.

So if you want to mark off 12½ feet, you can start from the ½-foot mark to the left of the zero, and then measure up to the 12th tick mark. This way the scale doesn't have a jumble of thousands of tiny subdividing tick marks.

You want your map to be as large as possible, yet still fit on the paper. Look at your longer measurements and estimate the length of the longest edge of your property; then look at your architect's scale and decide which scale to use so that the longest edge will fit. It's a good practice to have the top of your map oriented at least roughly toward north.

## DRAW YOUR MAP

Place a dot on the paper to represent your first anchor point. Then use the architect's scale to measure the distance to the second anchor point, and place a dot on the paper for that.

Choose the next point to add to the map (it's best to go in the same order in which you measured, in case any of these points become anchors later). Using the architect's scale, set your compass to the distance from the first anchor point to this point.

Center the compass on the dot for the first anchor point and strike an arc about where you estimate the new point will land. Then set the compass for the distance from the second anchor point to the new point; center it on the second anchor point and strike another arc — the arcs intersect at the location of the new point (Figure D).

**D**

Continue to draw your map, using the compass and measurements to triangulate each point. You might find that your compass won't extend to the longer measurements, even with an extender bar. If so, you can make a temporary compass by using a long strip of cardboard (try cutting a long strip off a file folder). Use a pushpin and push it through one end of the cardboard to use as the compass point. Use a second pushpin to make a hole at the distance you need; then put a pencil lead into that hole and use it to strike the arc (Figure E).

Draw everything in with pencil, and then ink over your lines to make them permanent. Use colored pencils, markers, or paints to add color shading to your map (Figure F).

Add a map legend with a rendering of the scale and a compass rose. Here's a quick hack to find north if you don't have a property line that runs north/south: print out the Google map of your neighborhood. North is always straight up the page, so use a protractor to find the angle between north and your property line, and then use that same angle (flipped over) to draw a line pointing north on your paper for your compass rose (Figure G).

Since I use my maps to plan changes I want to make, I like to draw in only permanent structures and trees and shrubs that I know will stay, and use tracing paper or vellum overlays to show movable structures or garden layouts (Figure H). You can also scan your map to use it on the computer or to create paper copies to work with.

**E**

*Always remember:* the map is not the territory! It's OK to sketch new ideas on paper of course, but when you're creating your planning overlays you should go out onto the property and do some more observing and measuring and then use that information to draw the features on your planning map.

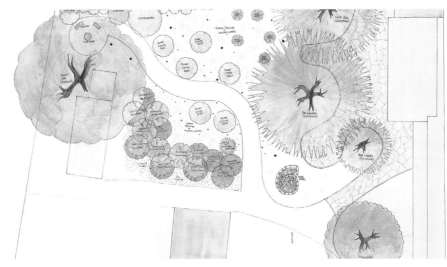

F

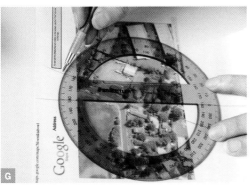

G

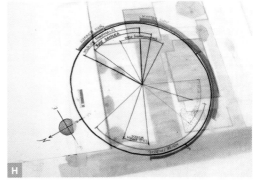

H

## ADD CONTOURS TO YOUR LOT WITH AN A-FRAME LEVEL

The A-frame level is a simple tool that's been used for centuries to create or mark level contours of the land. They were even used in the construction of the pyramids! You can make an A-frame level from simple materials that you probably have handy.

Lash the 2 longer sticks together at one end to form the sides of a long triangle. Then lash the smaller stick in place as a crossbar, completing the A shape. Make sure you lash everything tightly; the parts of the A-frame shouldn't move in relation to each other.

Tie one end of the string to the apex of your A-frame, and the other end to a weight (like a small

## TOOLS AND MATERIALS

Sticks or 1×2 lumber, in lengths of 2' (1) and 5' (2)
Heavy string or twine **for lashing sticks together.**
**You can also use nuts and bolts, if using lumber.**
Plumb bob, rock, or other weight **on a string**
Pencil
Permanent marker
Survey flags
Ruler

rock). The weight should hang about 4" below the crossbar (Figure I, following page).

Now you need to calibrate the level by making

marks on the crossbar. You'll make 2 temporary marks with a pencil, then a permanent mark with a marker.

Hold the A-frame up so that the string and weight swing freely and the string falls right next to the crossbar. Do this on a slight slope, with one leg upslope and the other downslope. Use survey flags to carefully mark where each leg of your A-frame meets the ground. Then use a pencil to make a temporary mark where the string falls at the crossbar (Figure J).

Flip the A-frame so that the legs are reversed — the side that was downslope before is now upslope, and vice versa. Make sure the legs meet the ground at the same points you marked with the survey flags. Use the pencil to make another temporary mark where the string falls at the crossbar now.

Now use your ruler to measure the halfway point between the 2 temporary marks on the crossbar, and mark this halfway point with a permanent marker.

Congratulations, your leveling tool is complete! When the string exactly crosses the permanent mark, it indicates that the feet of the frame are level with each other.

The A-frame level is especially good for marking

contours on the land. You've seen contours represented by lines on a topographical map — they're the lines of equal elevation.

To use your level to mark a contour on a slope, set one leg (A) of the level on a starting point and mark it with a survey flag. Keeping leg A in place, swing the other leg (B) around until it also stands on the ground and the string exactly crosses the center point.

To do this, you'll need to move leg B upslope and downslope until the 2 legs are level, as indicated by the string (but don't move leg A while doing it). Mark the position of leg B with a second survey flag.

Now pick up the A-frame and move it along so that leg A is at the flag you just placed for leg B; use this as the new pivot point, and again move leg B until the string falls along the level point. Mark the new point at leg B with a third flag. Repeat until you have your entire contour marked. When you finish, the flags mark the contour of the land (Figure K).

You can dig a swale or build a terrace along this contour. If you're placing a fence, you can count the spaces between the flags and multiply times the distance between the A-frame's legs, and you'll have the length of fencing you'll need to use.

## MAKE AND USE A WATER LEVEL

While the A-frame level is useful for laying out long runs of contour, the water level is useful for leveling the soil, marking points of equal elevation, and measuring short rises in elevation. The water level also works where there's no line of sight (around a corner, for example).

Constructing the water level is easy, but you'll want the help of a friend. First, set up a siphon with the tubing to get water flowing through it; make sure there are no air bubbles. Raise both ends of the tubing when it's full of water, then lower one end to allow a little water to escape. You want 12"–18" of empty tubing at each end, and a solid column of water in between.

With your friend's help, tape the tubing to the yardsticks. Make sure both yardsticks are oriented the same way, with the 1" mark at the top (Figure L). You can put the ends of the tubing even with the top of your yardsticks, but there's no need to get them exact (it's not the tubing you'll measure, it's the water level inside it). Leave a few inches untaped at the bottom, so that the stick touches the ground without interference from the tubing.

Here's how the level works: the water in one side of the tubing will always be level with the water in the other side. If you set each yardstick at a point, and the water measures the same height in both, then the yardsticks are standing on level points. If the measurements differ, subtract one from the other: the difference is the amount of rise between the 2 points (Figure M). The water will form a con-cave meniscus in the tube; for best results, always measure along the bottom of this meniscus.

How do you use a water level? Suppose you want to construct a wall with a top that's exactly level, and you've already driven posts into the ground along the path the wall will take. Hold (or tape) one of the yardsticks to the first post so that the water level is at the desired height of the top of your wall. Then take the other yardstick to each of the other posts, and mark each post where the water level equals that of the first post — that's the height of your wall. Then you can tie string between each point to mark the intended height of the entire wall.

Photograph by Terrie Miller (Figure L)

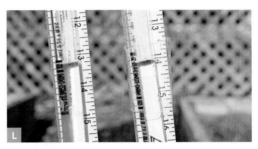

Terrie Miller lives in Northern California and publishes permie.net, a website about permaculture design and sustainable living.

NO LONGER WILL WE PASS GAS ONTO THE NEXT GENERATION.

# THE TIME IS NOW FOR ...

# ENERGY

# INDEPENDENCE!

THE *D.O.E.** CAN'T WAIT TO CATCH A *WHIFF* OF THIS *RESEARCH.*

*The D.O.E. (Department of Energy) is responsible for developing research into sustainable alternative fuel sources — including biological sources of power.

# HOWTOONS

I'VE **SEEN** THE NEWS STORIES.

The New York Times

# ICE SHELF COLLAPSES, PLANET WARMING

I'VE **LISTENED** TO THE **TV** PUNDITS FULL OF THEIR **HOT AIR**.

AND **I** CAN ONLY **CONCLUDE** THAT ADULTS JUST DON'T **"GET"** THIS **CLIMATE** AND **ENERGY PROBLEM**.

WE **NEED** RADICAL SOLUTIONS. IT'S TIME TO TAKE MATTERS INTO MY **OWN** HANDS.

**METHANE** IS A GREENHOUSE GAS WITH **20** TIMES THE POTENCY OF $CO_2$.

**BUT** IF I CAN CAPTURE IT, I CAN **USE IT** AS A BIO-FUEL TO CREATE ENERGY.

NOW ALL I NEED IS A **SPARK** TO GET THIS **ENERGY REVOLUTION** STARTED.

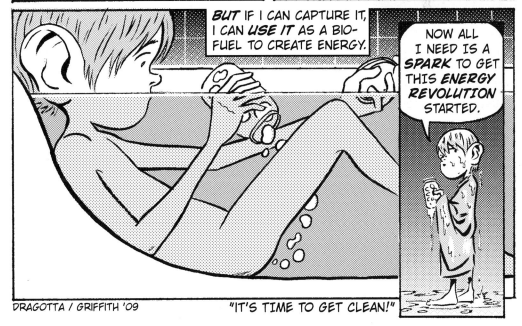

DRAGOTTA / GRIFFITH '09

"IT'S TIME TO GET CLEAN!"

# To the Bat Cave!

**The Scenario:** You and a friend, both experienced campers, are out for a wilderness weekend in one of your favorite desert areas when, around sunset you see a large, densely concentrated, directional swarm of bats sweeping low across the landscape. Intrigued by the sighting, you hike back along their flight path until you come upon the entrance to a cave from which they emanate — a cave that, as far as you know, is unknown. And the lack of any signs of human activity around the entrance seems to confirm that.

Not wanting to pass up this once-in-a-lifetime opportunity for some genuine adventure, you convince your friend to at least go in a little way to explore the cave. So, gearing up with packs and flashlights, you tie a guide rope to a bush near the entrance and slip through the narrow opening to have a look inside. The cave quickly widens, and once you get past the odor of the bats, you're both astonished by the undisturbed beauty of the structure. You venture deeper into this pristine geo-world until the rope runs out, but neither of you is keen to stop now. So you agree to make directional markers along the way in order to explore farther — marks cut into the cave floor or walls, piles of stones, bits of fabric; whatever they are, assume that they stick.

**The Challenge:** Taking many turns through naturally formed tunnels and chambers, you're both so involved that you finally realize that neither of you has been trail-marking for some time. The flashlights make everything seem bright, but when they're off it is pitch black. A little back-tracking doesn't find your last markers, and you realize you're most definitely lost. You hear the chittering of a few remaining bats and the dripping of water, but other than that and your breathing, the cave is deathly silent. So, aside from resisting the urge to panic, what do you do now?

**What You Have:** Two sturdy, aluminum-frame overnight backpacks, two canteens of water, some protein bars and other durable foods, two flashlights with extra batteries, a Swiss Army knife or Leatherman tool, a strong, flexible 3-foot wire saw with split-ring finger-handles on both ends, some waterproof matches, a compass, a cellphone (no, you don't get a signal down here), and a GPS locator (also no signal). Besides your hiking boots, you each have waterproof nylon rain gear and a nice warm jacket. So, can you find your way out, or have you truly reached the end of your rope?

Send a detailed description of your MakeShift solution with sketches and/or photos to makeshift@makezine.com by Aug. 21, 2009. If duplicate solutions are submitted, the winner will be determined by the quality of the explanation and presentation. The most plausible and most creative solutions will each win a MAKE T-shirt and a *MAKE Pocket Ref*. Think positive and include your shirt size and contact information with your solution. Good luck! For readers' solutions to previous MakeShift challenges, visit makezine.com/makeshift.

And the next MakeShift challenge could be yours! That's right, we're throwing open the doors and offering you the chance to create your own MakeShift to challenge the world. Just submit an original scenario in the familiar format — the challenge, what you've got, etc. — with some ideas of how you think it should be solved. The winning scenario will not only be published right here but also earn you a $50 gift certificate for the Maker Shed. The deadline is Aug. 21, 2009, so get out there and start looking for trouble!

Lee David Zlotoff is a writer/producer/director among whose numerous credits is creator of *MacGyver*. He is also president of Custom Image Concepts (customimageconcepts.com).

Photograph by Jen Siska

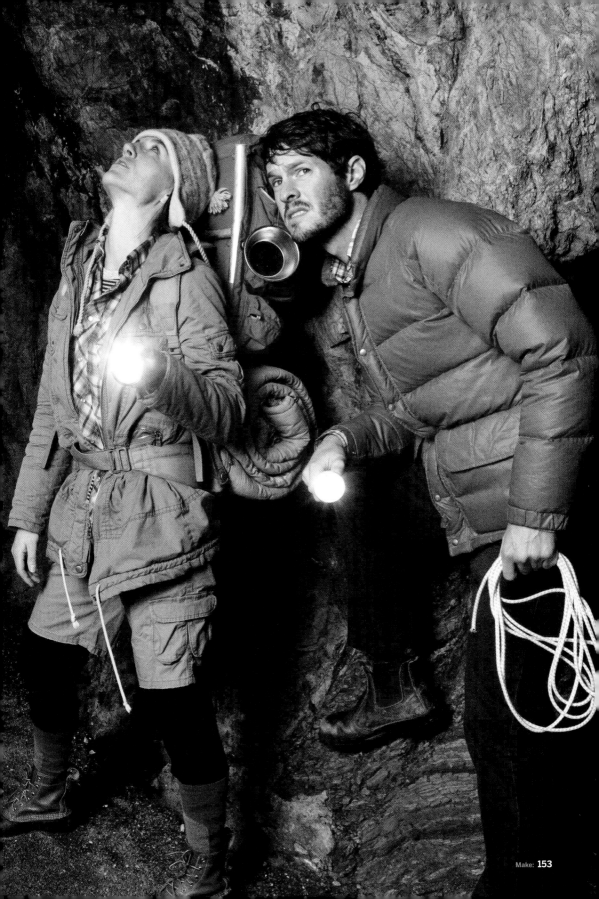

# Maker Family Portrait

A family of inventive blacksmiths in Indonesia supply the locals with just about everything they need.

>> At the far western tip of Papua, on the outskirts of the town of Sorong, is a place called Tempat Garam ("salt-making place"). The Mombrasar family (pictured at right) of blacksmiths have their shop there. They build boats, make any and all kinds of tools, and invent labor-saving devices.

### Chainsaw-Powered Sago Grinder
Yohanis Mombrasar showed me one of the family's inventions, a chainsaw-powered sago grinder (Figure A). The local staple is sago palm starch. The sago palm grows in dense stands in freshwater swamps just behind a barrier beach. The pith of the trunk is composed of starch and fibers.

Big chainsaws are plentiful here because of the timber industry. The area has valuable hardwoods sought by Malaysian Chinese traders.

The traditional method of making sago starch is to fell a sago log and pound the insides with wooden hammers until the starch grains are separated from the fibers. (I've read that even that way, it's one-tenth as much labor as rice cultivation.) With a power tool like this, it would take very little time to process large quantities.

### The Product Line
*Pandai besi* means "blacksmith" in the Bahasa Indonesia language. Just like our word, it literally means "iron pounder."

This sign (Figure B) shows some of the things that someone in this family is ready to make at any time, including axes, a huge variety of knives and machetes, spear points, sickles, chisels, and all kinds of hardware.

The local stores carry mass-produced machetes and sickles like we have, but no one wants them. The local people appreciate a finely crafted steel tool made exactly to suit the work they do.

ALL IN THE FAMILY: (L–R) Vincent Ambrar, Fredison Mombrasar, Andreas Mombrasar, Elisabet Dimara, Delilah Mombrasar. The kids were moving around too much for me to get their names straight.

This price list (Figure C) shows how much the family's major products cost and how many they can make in a month. It's an impressively well-organized operation.

### The Forge
It was their day off, but Elisabet Dimara and Andreas Mombrasar kindly offered to show me how their forge works. It's a very sociable operation.

Elisabet sits on the throne and works the bellows (Figure D). It's a piston pump made from two sections of water pipe and some wooden piston plungers. The gasket material is very soft and hangs down on the upstroke, allowing air to pass around it. It looks like a soft foam. I've also seen gaskets made from many layers of woven plastic bags.

*Tuyère* is the English word, from French, for the pipe that blows air into a forge. Some call it a *tweer*. Theirs is made exactly as it appears; two small pipes,

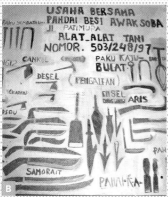

no valves. The iron rod in the middle is only there to set the spacing between the two nozzles (Figure E).

The nozzles blow air through a hole in the side of a vertical slab of tile or stone. The side blast probably reduces the problems with clinkers clogging the tweer. That's a chronic problem with American forges, which blow air from below.

Andreas lit some wood shavings and piled charcoal on them. Elisabet worked the pistons, and after a few strokes the forge was roaring (Figure F).

## Coconut Grater

Another fantastic Mombrasar family invention is a coconut-grating attachment for a hand-cranked knife sharpener (Figure G). You hold half a coconut against the spinning cutter and put a bowl underneath to catch the grated coconut.

They also make really nice traditional coconut-grating tools. I bought a couple of those. The ornate wooden thing in the foreground is a spear gun with a trigger made from a nail.

## Transportation

Here's one of the family's fishing canoes (Figure H). The design is from Ambai Island. I didn't ask if they have relatives there, but that island has a strong blacksmithing tradition.

The crossbeams are exactly 1 meter apart, which is how much room a paddler needs. The outrigger log is a branch of hibiscus, a very light wood.

They're also repairing this huge dugout canoe (Figure I). The local water taxis are just like this. I think it's made from a smaller log, which is spread open with steam, but I haven't seen it done.

---

Tim Anderson (mit.edu/robot) is the founder of Z Corp. See a hundred more of his projects at instructables.com.

Photography by Tim Anderson

**Wireless signal detector for the paranoid, a fire-powered soak, iPhone hacks, and tales of sustainability.**

# TOOLBOX

### Leatherman Wave
$90 leatherman.com

I just got a Leatherman Wave multi-tool, and to say I'm in love is an understatement. In the three days I've had it, it has left my side for no more than 15 minutes at a time. In addition to the two outer blades (serrated and normal) that can be deployed in seconds using only one hand, there's a diamond-coated file and a saw.

Inside, you find not only the standard pliers (which are beautiful all by themselves) but also two bit drivers (large and small), a bottle opener, a ruler, and more.

Remember those flimsy Swiss Army scissors that can barely cut paper? The Wave's scissors are far better; they have a real handle, not to mention that you can tell just by their look that the Swiss Army blades can't even compare.

No matter which tool you're using on this amazing 17-function knife, it feels great in your hand. As an added bonus, you can purchase a bit kit, which is full of different-sized flat, Phillips, and Torx bits.

Within two hours of getting my Wave, I'd already used it to fix my boot. Earlier today, I used it to help me build an iPod charger in an Altoids tin, from Volume 07 of MAKE.

—*Adam Zeloof*

## TSX300 Portable Two-Way Radio

$40 trisquare.us

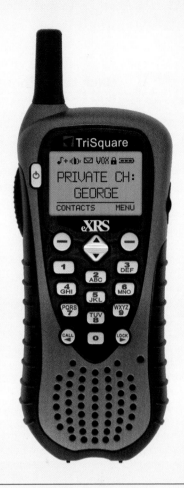

Ever tried to stay in touch with friends at a mall or sporting event using Family Radio Service or GMRS radios? Competing with other users can be a challenge, even with 21 channels and so-called privacy codes.

Switch to "eXtreme Radio Service" (eXRS) portables by Tri-Square Inc. to eliminate interference and ensure voice privacy. eXRS radios use digital spread spectrum (DSS) modulation, converting speech into a digital data stream transmitted over constantly changing 900MHz frequencies, to eliminate eavesdropping by all except eXRS radios you authorize.

The top-of-the-line TSX300 includes text messaging, multiple virtual channels and user groups, and a NOAA weather receiver. Initial setup simply requires programming each radio with a user-selected channel code, with wireless cloning possible.

I found the TSX300 to be superior to FRS/GMRS units in nearly every respect. A slight reduction in range was more than offset by communications privacy and freedom from interference. It sells for about $40 including rechargeable battery, charger, earpiece speaker/mic, and belt clip. No license is required, and they're FCC-approved for both personal and business communications.
—*L. Abraham Smith*

## Hidden Camera Wireless Signal and Wi-fi Detector

$11 dealextreme.com

Do you sometimes feel you're being watched? Listened to? Are you quite certain you're alone? Nobody will think you're odd if you insist on using code words or ROT13 when speaking in public, but in the safety of your own home, it's a bit excessive. That's why DealExtreme's Hidden Camera Wireless Signal and Wi-fi Detector is such a boon to the hopelessly paranoid and spies on a budget.

The unit is a tiny handheld device, smaller than most cellphones and not even the size of six isotope disks of polonium. A single button and two LEDs adorn the front of the detector. Press the button and the red LED lights up to indicate that the unit is activated. When the unit detects frequencies in the 100MHz to 3GHz range, the blue LED will blink. The more activity it detects, the faster the LED blinks.

The unit cannot, alas, distinguish between good and evil. To get a decent reading, you'll have to turn off all your wireless devices. Then you can walk the house, antenna outstretched, searching for bugs.

I couldn't find any bugs not of my own making, but for the wireless devices I planted myself, the detector worked flawlessly. It was easy to hone in on a device just by watching the frequency at which the LED blinked. The detector's price is $11, and that includes shipping from Hong Kong. It's a bargain if you have even a passing suspicion that the walls might have ears.
—*Tom Owad*

## Hack Your iPhone and iPod Touch

**iPhone Hacks** by David Jurick, Adam Stolarz, and Damien Stolarz

$35 Make: Books

*Editor's Note: Instead of a review of this book, we thought we'd let the book speak for itself and offer you an excerpt of four helpful hardware hacks.*

### Hacking the Wire

The technical problem that normally prevents most headphones from connecting to the iPhone is the plastic sheath at the base of the plug. The girth of this plastic piece is too wide to fit in the narrow recess surrounding the iPhone's headset port. If you compare the plug of the iPhone headset to that of a standard set of headphones, you'll notice the difference in thickness. To get the 3.5mm plug of almost any standard pair of headphones to fit in your iPhone, you'll need to shave off about 5mm of the plastic sheath.

### Noise-Cancelling Headphones

What you may not know is that over-the-ear noise-cancelling headphones work just fine over the headphones that come with your iPhone. If you

own an original iPhone you'll get the benefit of noise-cancellation, without having to buy an adapter.

### Connect the iPhone to a Car

One problem with iPod integration in cars is that iPhone integration isn't perfect. For one thing, the iPhone 3G won't charge via many of the iPod integration systems, because they provide 12V to the FireWire pins instead of 5V. [Here are] two adapters that solve this problem: one by Scosche (scosche.com/products/productID/1667) and the other by CableJive (cablejive.com/chargeconverter.html).

### Controlling Your iPhone/iPod Touch in the Car

The name of the game in driver safety is keeping your eyes on the road, hands on the wheel. Scosche also makes a clever device for controlling your iPod, from your steering wheel, which also happens to work for the iPod touch and iPhone (scosche.com/products/sfID1/210/sfID2/324).

---

## *Bob's Knee*

**Directed by Michael Attie**

Free online at vimeo.com/4045734

After having his own knee problems, inventor Bob Schneeveis started thinking about the way humans walk. "How do we walk?" he asked, and then admitted that he didn't know. "The way to really know how something works is to make a model of it." And so he created machine models of a walking man. His work culminated in the hand-built machine he calls RoboChariot, which roams the grounds of Maker Faire every year.

Schneeveis is the subject of a short documentary film by Stanford film student Michael Attie called *Bob's Knee*. The film, shot in 16mm black and white, is hand-crafted as well. "I used a Bolex, which is a spring-wound 16mm camera," says Attie. "My model was from the 60s, but the design hasn't changed since the 30s."

Last year, Schneeveis had his walking man fitted with a mask of Arnold Schwarzenegger. This year's model will feature President Barack Obama.

—*Dale Dougherty*

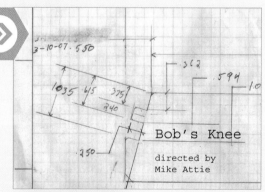

📹 **Michael Attie's site:**
mikeattie.com/Mike_Attie/Bobs_Knee.html

## World's First Embedded Power Controller

**Cypress PSoC FirstTouch Starter Kit**

$70 cypress.com

As someone whose microcontroller experience mostly centers around Arduino, the Cypress PSoC was very new to me. It's a chip that's part micro-controller, but the chip itself is reprogrammable; you reconfigure the circuits inside the chip, on the fly.

Want to turn a couple of pins into an I2C bus? Need to have a serial conversation over USB? No problem. And in addition to programmable logic, the chip also has a bunch of op-amps you can reconfig-ure on the fly. Want to generate some DTMF tones for wardialing? You can do that, too.

I got started with the PSoC FirstTouch kit. It comes with two other boards with chips and sen-sors on them, and a dongle that plugs into a USB port. Two of the devices (the dongle and one board) include RF chips for wireless communications.

If you want to build a PSoC-based project from the ground up, a PSoC DIP chip (suitable for a bread-board) runs less than $10 in single-unit quantities. The $25 PSoC MiniProg USB programmer (Digi-Key part #CY3217) is all you need to program the chip.

All the software needed for programming the PSoC is included for free (and available as a free download), but it has two downsides: it runs only on Windows, and the C compiler operates in a reduced-functionality mode ($1,995 for the full version).

Programming the PSoC is very different from what I'm used to: yes, there are some bits of C code at the heart of most projects, but the design software is visual: you'll drag, drop, and connect analog and digital modules to define the way the chip behaves.

*—Brian Jepson*

## Like Clockwork

**Chronulator 2.0 Clock Kit**

$49 sharebrained.com/chronulator

Alternative time-telling devices are compelling, but only if they're easy to read. Building a kit can be sat-isfying, but only if it leaves room for creativity. The Chronulator 2.0 clock kit fits the bill on both counts.

Solder it together, and you've got a microcontroller-based clock that converts time to current, displaying hours and minutes on two analog panel meters. Print out the supplied clock face templates, or customize the meters. No housing is supplied, and this is where it gets interesting. I mounted my panels into a cigar box, and put the circuit board on top, its exposed wires lending to the retro-tech design. (I also consid-ered using an old Mac G4 Cube case, or mounting it naked to the wall in a PanaVise.)

Based on an Arduino-compatible Atmel AT-mega168 chip, the Chronulator lets you download and modify the source code, connect to a computer via a USB-to-serial adapter, and display any kind of data. Would it be crass to have a "Number of People in My Facebook Friend Requests Purgatory" meter?

The Chronulator runs on a minimum of 1.8 volts and about 200 milliamps, so it'd be easy to power it from a small solar cell and a super capacitor. Then we'd have green time!

*—John Edgar Park*

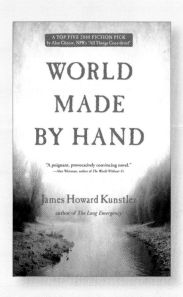

## « In the Future: Back to Basics

***World Made By Hand* by James Howard Kunstler**
**$14 Grove Press**

In this sweet and sad novel, the population of the United States (and most likely, the world) has been decimated by an energy shortage, starvation, plagues, terrorism, and global warming. The story takes place in an unspecified time in the future (I'm guessing around 2025 or so).

The story is told by Robert Earle, a former software executive. Now he's a hand-tool-using carpenter living in upstate New York. The electricity comes on every couple of weeks for a few minutes. When that happens, nothing's on the radio but hysterical religious talk. Rumors of goings-on in the rest of the world are vague.

While life is lawless and harsh, it's not without charms. Local communities are active and productive. Neighbors know and help each other. People grow and trade their own produce and livestock, and meals are tasty — lots of buttery cornbread, eggs, chicken, steaks, vegetables, and fish. They get together and play music, and because people aren't in their living rooms watching TV, they attend live shows.

As a budding urban homesteader, I found this way of life fascinating. No one can predict the future, and I doubt ours will be much like the one depicted here, but I think it's possible that Kunstler has come closer to showing us what might be in store than anyone else.

—*Mark Frauenfelder*

## « Simple, Clean, and Green

***Clean: The Humble Art of Zen-Cleansing* by Michael de Jong**
**$8 Joost Eiffers Books**

I've carried around this book for at least a year. At times, I've felt like a clean-freak Johnny Appleseed, whipping it out and offering up easy, green cleaning solutions to anyone who mentioned they had a dirty problem. And people keep coming back to borrow the book or look up another cleaning technique in the alphabetical index.

The list of ingredients includes just 5 common items: lemon, salt, baking soda, vinegar, and borax. The charts and instructions are elegantly simple: "Soak your rusty tools in white vinegar for a few hours or overnight." And de Jong's snippets of wisdom are to the point: "Fix it, change it, clean it, make it better, or get rid of it."

At home, we've used it over and over again, and I've been amazed at how it's helped get my copper teapot clean and de-tarnish a silver trinket, all while keeping the entire family entertained.

With strong-smelling and potentially harmful commercial products, there's no way I'd want my kids to help polish copper or brass. But using some salt and a half a lemon rubbed in circles is fun, smells nice, and is a bit messy — perfect for my 6-year-old to help with. Plus, the results are nothing short of magic.

Even more magical was the neat way to get the tarnish off silver plate. Who knew aluminum foil, boiling water, and baking soda could make tarnish literally jump?

—*Shawn Connally*

## « Modern City Living, Perhaps

**The Urban Homestead: Your Guide to Self-Sufficient Living in the Heart of the City** by Kelly Coyne and Erik Knutzen
**$17** Process Books

This is a delightfully readable guide to front- and backyard vegetable gardening, food foraging and preserving, and other useful skills for anyone interested in taking an active role in growing the food they eat. I learned about composting, self-watering containers, mulching, raised bed gardens, and raising chickens by reading this info-dense book.

Unlike many self-sufficiency books, this one isn't unrealistic, preachy, or dogmatic. Instead, it's honest and often humorous. Coyne and Knutzen are wonderfully lucid and accessible writers (*Knutzen wrote a drip irrigation how-to on page 72*). They also walk the walk — I visited their Los Angeles home, touring their abundant vegetable gardens and henhouse filled with clownish chickens — plus they run a terrific blog at homegrownevolution.com.     —MF

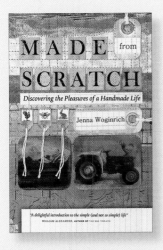

## « Stabs at Meaningful Living

**Made from Scratch: Discovering the Pleasures of a Handmade Life** by Jenna Woginrich
**$21** Storey Publishing

Jenna Woginrich is a young web designer who wanted to meaningfully participate in the systems that keep her alive and well. Her book is a humorous and useful account of her attempts to raise chickens, grow vegetables, keep bees, raise rabbits, play mountain music, and preserve food. She's not always successful (varmints tore her beehive apart, for instance) but she doesn't let mishaps discourage her from experimenting with new ways to become more self-sufficient. The end of the book has an appendix with resources for getting started in modern homesteading. You can follow her continuing experiences on her blog at coldantlerfarm.blogspot.com.     —MF

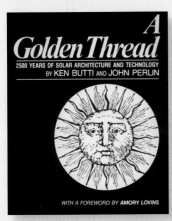

## « Solar Tech History

**A Golden Thread: 2,500 Years of Solar Architecture and Technology** by Ken Butti and John Perlin
**Out of print** Cheshire Books, Van Nostrand Reinhold

I'll never forget the first image I saw when I picked up this book a few years ago — a vista of Los Angeles circa 1900 looking shockingly akin to a rural Swiss village, a number of roofs scattered with mysterious boxes soon explained to be solar water heaters.

In this magnificent book on the history of solar architecture and technology dating as far back as ancient Greece, the authors offer beautiful, often humbling technologies, city plans, and outlandish ideas that were somehow replaced, abandoned, or just plain forgotten.

More than just a fun read on the focusing of resources, *A Golden Thread* boldly challenges misconceptions that solar power is still in its developmental stages, yet leaves you feeling like this is just the beginning of something great.     —Meara O'Reilly

## Chofu Wood-Fired Hot Tub

$800 islandhottub.com

The Chofu is a very simple, and very beautiful, DIY wood-fired hot tub heater. It looks like a potbellied stove with an Eastern aesthetic, and it comes from Japan. Any container that will hold water can be the hot tub (metal stock tanks are a great choice).

The stove is connected to the tub by two openings. The lower opening allows cold water to fill the stainless steel water jacket that makes up the Chofu's round sides and top. As the water is heated by the fire, hot water rises to the top and pours out the upper opening into the tub, and colder water from the tank is drawn back in. This is the power of a thermosiphon; no pump is needed. The Chofu naturally circulates the water, letting you have a hot soak completely off-grid.

We got our Chofu in the mail, and setting it up was a very easy project. In fact, the hardest part is connecting the pieces of stovepipe together. It can all be done in under an hour.

To keep things eco, try burning pressed wood logs made from industry sawdust byproduct, and be sure to choose logs free from adhesives. Stoke the fire every 45 minutes, stir the water as well, and in about 4 hours, a 250-gallon tank will be close to 104°F.

The Chofu setup is very open to mods: employ a lid to keep your tub cleaner, longer, without the use of chemicals, or insulate the tub to increase efficiency. It also appeals to the DIYer who's budget-minded. While the stove ($800 including shipping) and tank ($200 locally) will certainly set you back, it's a fraction of the cost of a new manufactured hot tub, and the Chofu will never cost you a penny in electricity.

It's a must for the modern homestead.

—Brookelynn Morris

---

## Tricks of the Trade By Tim Lillis

*Keep TP on hand.*

Have trouble keeping track of your TP stock? Looking for a fun and functional way to liven up the bathroom? Reuse a wrapping paper tube or other long cylinder!

First, wrap the top end in tape. This saves the tube from wear and tear and allows you to effortlessly add rolls of toilet paper to the stack.

Next, glue some cardboard or similar material at the bottom to create a lip that the TP rolls will rest on.

Now you have an easily accessible stack of toilet paper that allows you to take a quick visual inventory at any time.

Have a trick of the trade? Send it to tricks@makezine.com.

# Get Your Fix

**Free** fixya.com

The radio on our Honda Civic Hybrid stopped working while we lent it out last year. The display just prompted "Code," which we didn't have since it was a salvage car we'd bought from a restorer. Our (replacement) owner's manual informed us that this was an anti-theft feature that activated when the car's power disconnected. Two garages I called told me that I had to take the car in and "have the radio pulled," but after searching various online forums, I found a better solution. Thanks, honda carforum.com member honda-guy!

More recently, I learned about FixYa, a website dedicated to sharing troubleshooting tips for all products. The Honda radio code trick is in there (it's the first thing I checked), as are countless other nuggets, all organized clearly and ranked by other users. I've seen other sites attempt a similar function without reaching the critical mass needed to succeed, but FixYa seems to have finally done it, becoming the place people know to go for this kind of information. I have since "paid it forward" by posting a Netgear wireless router trick I learned from a patient and cheerful man on Netgear's technical support line. Meanwhile, we've had no trouble with the Civic, and the only mystery is why its power was disconnected in the first place.　　　*—Paul Spinrad*

## Game Creation for Everyone

**Multimedia Fusion 2**
**$120** clickteam.com

If you can confidently surf the web, edit graphics on a computer, or use a spreadsheet, you can probably use Multimedia Fusion (MMF) to write your own programs and games. And you won't have to learn a cryptic computer language either. Best of all, it's fun!

MMF is a computer program that lets you easily create Windows programs. The secret is that a team of programmers have done a lot of the heavy lifting behind the scenes, so instead of writing code, you can "drag and drop" and use menus to create your program. My 9-year-old son has created his own games using MMF, and together we made a simple CD player in about 5 minutes. I use MMF to write computer programs for myself as well as the museum I work for, and the results have been outstanding.

My favorite thing about MMF is that it makes me look smarter than I really am. It lets me focus on what I want my program to *do*, rather than the actual coding. It's like having access to a whole team of software developers.

The only drawback is that there's no Mac version. Still, there's a free trial version at clickteam.com, and it's very helpful to talk to other users in the forums.　　　*—Dave Stroud*

**Dave Stroud** programs for fun and works for children's museums ... for fun.

**Tom Owad** is a Macintosh consultant and editor of applefritter.com. He is the author of *Apple I Replica Creation*.

**John Edgar Park** is the host of the Maker Workshop on *Make:* television (makezine.tv).

**Brookelynn Morris** says, "It's impossible to be angry while having a hot soak." Her first book, *Feltique*, is the definitive guide to working with felt.

**Meara O'Reilly** is an intern at MAKE.

**L. Abraham Smith**, amateur radio licensee N3BAH, works on open source software, hardware, and Linux system administration every chance he gets when not practicing law.

**Adam Zeloof** lives in Central New Jersey and enjoys sailing, camping, geocaching, and of course, making.

**Brian Jepson** is an editor for O'Reilly Media Inc. and a co-founder of Providence Geeks. He likes to hack electronics, software, and gadgets.

Have you used something worth keeping in your toolbox? Let us know at toolbox@makezine.com.

# Reflections on an Illusion

An illusion that never ceases to fascinate is the mirror-produced *real image*. It can appear so real that you'll reach out to grasp it, but your fingers close only on thin air.

» Nineteenth-century books of science recreations often included the "phantom bouquet" (Figure A), which produced an upright real image of a bouquet of flowers hidden in a box below. The flowers had to be upside down because the concave mirror inverts the image.

You can duplicate this pretty effect with the concave side of a large shaving or cosmetic mirror, as shown in the old engraving in Figure A.

A somewhat more stable version uses a wooden box or stand enclosing the bouquet on all sides except the one facing the mirror. Paint the inside of the box matte black, or, better yet, line it with black plush or velvet cloth. Illuminate the bouquet, perhaps with a small desk lamp. The bouquet, and its image, will both be a distance $R$ from the mirror, where $R$ is the mirror's *radius of curvature*.

A nice variation is the "phantom light bulb." A light socket is fastened to the top of the box, and a second one is fastened upside down inside the box below it. Only the latter is powered. When the lower bulb is lit, its image appears in the socket above, if everything has been well aligned. If you use a clear bulb, get two of them. Put one bulb in each socket. Turn on the power, and the bulb above seems to be glowing. You can then unscrew the bulb and remove it, yet its phantom remains.

If you use a small-diameter mirror, the object should be relatively small. Use an old Christmas lamp and socket, or a decorative candelabra lamp. Frosted bulbs work best. Electrical and electronics supply stores have sockets that can be mounted on a flat surface.

One problem remains. You need a large concave mirror for the most effective presentation, and shaving or cosmetic mirrors are a bit small. Larger mirrors can be purchased at scientific supply houses, but are a bit pricey.

## An Educational Toy and A Work of Art

For less than $50 you can buy such mirrors, packaged as a device that produces another neat version of this illusion. The Mirage is made by Opti-Gone (optigone.com) in two versions: with 9" or 22" mirror diameter.

Assembled, it looks something like the conventional flying saucer, complete with a hole in the top from which you'd expect tiny aliens to emerge (Figure B). It's sold by science stores, museum shops, and science-supply sources.

Most people put this device on a table and display the illusion as a work of art (Figure C). But I'm sure MAKE readers will want to take it apart and see what else can be done with its parts.

The Mirage consists of two parabolic mirrors, the upper one having a large hole in its center. The mirrors fit together in a clamshell arrangement, separated exactly two *focal lengths* apart. When light from a distant object falls on a spherical mirror, the rays converge to a point called the focal point that is about $R/2$ from the mirror, where again $R$ is the mirror's *radius of curvature*. The convergence isn't perfect, and it's much better when the mirrors are parabolic. For mirrors this size, you don't notice the slight difference between spherical and parabolic.

The mirrors of the Mirage can be used separately. In a darkened room, let the light from an open window fall onto one mirror. An image of the window will appear near the mirror's focal point. You can only see it if the reflected rays from the mirror surface enter your eyes, so you have to look back toward the mirror surface. This is a bit tricky because your head gets in the way of some of the incoming light. Or, you can place a small sheet of paper at the image location, and see a reduced-size image of the window projected onto the paper.

Figure A from George M. Hopkins, *Experimental Science*, Munn & Co., 1890. Photography courtesy of Michael Levin, Opti-Gone. (B, C); and by Donald Simanek (D, E)

Concave Mirror, Phantom Bouquet.

**Fig. A: The phantom bouquet, from George M. Hopkins' *Experimental Science*, Munn & Co., 1890.**
**Fig. B: The Mirage optical illusion shows an image of a bolt and nut, as well as their reflection in the mirror below.**
**Fig. C: The Mirage shows another real image floating in space. Fig. D: Touching the real image (left) of your finger**
**(right). Fig. E: A real plastic frog (bottom) and its real image (top).**

You can also demonstrate the convergence of distant light to a focal point by using the sun as a distant source. A small tissue of paper placed at the focal point can be easily ignited.

⚠ **CAUTION: Don't let reflected sunlight fall into your eyes. Never look directly at the sun using any optical instrument, such as binoculars or a mirror. Children should always be supervised when doing experiments with bright light.**

Physicists call these images *real images* because convergent light really passes through them, and then diverges just as scattered light would diverge from a real object. Real images are distinguished from *virtual images*, such as those from a flat or convex mirror, where the light rays don't actually pass through the space where the image appears to be located.

## Touching the Real Image

Using only one mirror, place your finger at a distance of about *R* from the mirror's center. You should see a real image of your finger. With a little manipulation of its position you can create the illusion of your real finger just touching the real image of your finger (Figure D). Since you have two eyes, you see this illusion in three dimensions and can confirm the location and orientation of the real image in space. Note that the image is reversed in all ways: up/down, right/left, and top/bottom.

## The Phantom Reflects

Carefully place a small object at the very bottom of the lower mirror. Now assemble the clamshell with the upper mirror in place. Floating just above the hole in the upper mirror, you'll see a real image of the object inside (Figure E). Note that the image is rotated 180° about a vertical axis, with respect to the object, but the image is still right side up.

One illuminating demonstration isn't mentioned in the user's manual. Place the mirrors in the standard arrangement, with a small object inside. Its image appears just above the hole. Now shine a flashlight onto the image, but aimed so that the light actually

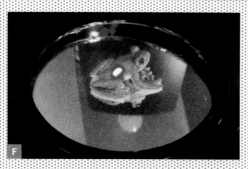

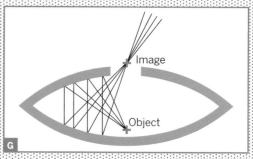

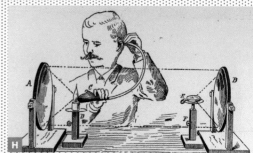

Fig. F: Aiming a laser pointer beam at the image of a frog causes a small spot on the image to be illuminated. Fig. G: Typical ray paths from object to image and beyond. Fig. H: Two concave mirrors can be used to demonstrate images formed by reflection — with light, heat, or sound. Here, sound from a ticking watch is focused onto a flame, causing it to flicker. Light from the flame forms an image on the watch.

passes down into the hole of the upper mirror. The image will display shadowing just as if it were a solid object being illuminated by the flashlight. This can be startling if done in a darkened room with the flashlight as the only source of light.

Aim a laser pointer beam at the image (again, at an angle such that the beam really passes through the mirror hole), and a small spot on the image will be illuminated by the laser beam (Figure F). You can even use a small, flat mirror as an object, then get a reflected image from the image of the mirror!

The ray diagram (Figure G) for this device reveals the reason. Room light falling from many directions through the hole illuminates the object lying on the bottom mirror. It scatters onto the top mirror, then reflects to the bottom mirror and back up through the hole. All rays from a particular point on the object finally converge to a particular point on the image, just above the hole, and then diverge just as scattered light would diverge from a solid object.

But what about the laser beam? Here an important principle of optics comes into play. The path of light through optical systems is reversible. Light from an object point may take a complicated path to reach the image point. But light from that image point can be directed back to reach the object point. When you send a laser beam through a point on the image, it reaches a point on the object, then scatters, and goes back through the optical system to the image point again, and proceeds as if it originated there.

## Other Demonstrations

» The Mirage comes with a small, pink plastic pig. Lay it on its side, and its image will appear to have accidentally fallen over. Now watch people, seeing this for the first time, reach for the pig to set it upright, and find nothing there.

» Place a coin in the center of the lower mirror. Put a sheet of flat glass over the upper hole. The coin image seems to be resting on the glass. Place a real coin on the glass, aligned perfectly. Tell your victim how real the image looks, then reach down and pick it up, leaving the real image there.

» A single Mirage mirror can be used in the vertical position to re-create the classic phantom bouquet and phantom light bulb illusions.

» Sound reflects in the same manner as light, so a mirror can be used to make a parabolic microphone. Put a microphone at its focal point. One mirror can be used as a sound transmitter and the other as a receiver, separated by large distances (Figure H).

Donald Simanek is emeritus professor of physics at Lock Haven University of Pennsylvania. He writes about science, pseudoscience, and humor at www.lhup.edu/~dsimanek.

Photograph and diagram by Donald Simanek; Figure H from Hopkins, 1890

# MAKER'S CALENDAR
Compiled by William Gurstelle

Our favorite events from around the world.

## Ingenuity Festival

July 10–12, Cleveland, Ohio

Ingenuity is a celebration of art and technology that involves the audience "as both spectator and participant." High tech firms and major colleges are represented alongside acclaimed artists, for a wide-ranging display of robotics, theater, interactive technologies, visual arts, film, music, and more. ingenuitycleveland.com

## ›› JUNE

### World Science Festival
June 10–14, New York City
With over 100,000 participants last year, this science fest boasts a variety of lectures, multimedia presentations, and discussions. worldsciencefestival.com

### Made In America
June 17–20, York County, PA.
Twenty factories, from Harley-Davidson motorcycles to Utz potato chips, throw open the doors and let visitors go behind the scenes. factorytours.org

### Brickworld
June 18–21, Wheeling, Ill.
A convention for grown-ups who still enjoy playing with Lego. Experience elaborate creations, workshops, presentations, and special challenges. brickworld.us

### Secret City Festival
June 19–20, Oak Ridge, Tenn.
Pays tribute to Oak Ridge's rich history as part of the Manhattan Project. It's the largest World War II reenactment in the Southeast and combines nuclear energy facility tours with entertainment and arts. secretcityfestival.com

## ›› JULY

### Darwin 2009 Festival
July 5–10, Cambridge, England
This year marks 200 years since Charles Darwin's birth and 150 since his earth-shaker, *On the Origin of Species*. Activities include discussions, workshops, performances, exhibitions, tours, and talks, including guest speaker Sir David Attenborough. www.darwin2009.cam.ac.uk

### SolarFest 2009
July 10–12, Tinmouth, Vt.
Musicians, dancers, and performance artists enhance this maker-centric event focused on solar power. Noted speakers hold dozens of workshops on renewable energy, green building, and sustainability topics. solarfest.org

### Table Mountain Star Party
July 23–25, Ellensburg, Wash.
The TMSP is an annual gathering of amateur astronomers who enjoy the great night-sky viewing that Table Mountain provides. Anyone is welcome to register and enjoy the experience. Celebrate 2009, the International Year of Astronomy! tmspa.com

## ›› AUGUST

### Sound of Light
Aug. 1–15, Gatineau, Québec
An international "pyromusical" competition (fireworks synchronized with music) set beside beautiful Leamy Lake. feux.qc.ca

### International Symposium on Electronic Arts (ISEA)
Aug. 23–Sept. 1, Belfast, Northern Ireland
The Inter-Society for the Electronic Arts promotes creative cross-pollination between the arts, sciences, and emerging technologies. Planned topics include interactive textiles, wireless sensor networks, wearable computers, and performance measurement in medicine and sports. isea2009.org

IMPORTANT: All times, dates, locations, and events are subject to change. Verify all information before making plans to attend.

*Know an event that should be included? Send it to events@makezine.com. Sorry, it's not possible to list all submitted events in the magazine, but they will be listed online.*

*If you attend one of these events, please tell us about it at forums.makezine.com.*

Photography by Janet Macoska

# AHA!  **Puzzle This** By Michael H. Pryor

**MAKE's favorite puzzles.** (When you're ready to check your answers, visit makezine.com/18/aha.)

## Letter Rip

Figure out what the following abbreviated phrases mean. I've filled in the first one to get you started:

0.  24 H in a D = 24 Hours in a Day
1.  26 L of the A
2.  7 D of the W
3.  7 W of the W
4.  12 S of the Z
5.  54 C in a P (W Js)
6.  13 S in the U S F
7.  18 H on a G C
8.  5 T on a F
9.  90 D in a R A
10. 3 B M (S H T R)
11. 32 is the T in D F at which W F

12. 3 W on a T
13. 12 M in a Y
14. 8 T on an O
15. 29 D in F in a L Y
16. 365 D in a Y
17. 13 L in a B D
18. 52 W in a Y
19. 9 L of a C
20. 60 M in an H
21. 23 P of C in the H B
22. 64 S on a C B
23. 1,000 Y in a M

Michael Pryor is the co-founder and president of Fog Creek Software. He runs a technical interview site at techinterview.org.

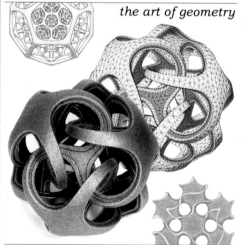

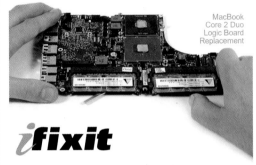

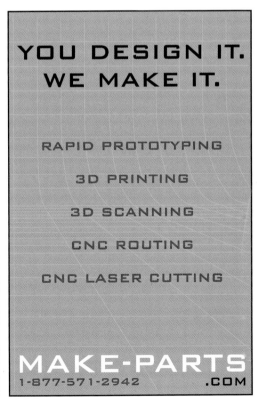

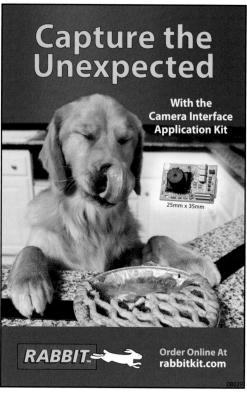

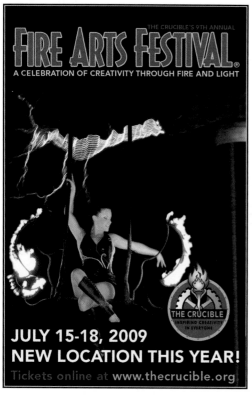

# MAKE MONEY | Salt & Pepper Shakers By Tom Parker

Sometimes it costs more to buy it than to make it from the money itself.

## $80.00
**Metal salt & pepper shakers from bestwishes.net.**

## ⬆ $20.42
**Salt & pepper shakers made from two rolls of drilled-out quarters, a scrap of ½" copper pipe, and a nylon plug.**

Photograph by Tom Parker

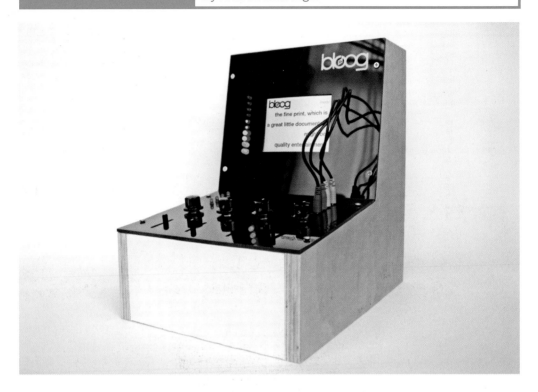

■ **You know when companies combine two words** to make a brand name? Like Verizon, or Go-Gurt? There's a term for this, *portmanteau*, which I just learned from Wikipedia, whose name is itself a portmanteau. I'm surprised I didn't know the term, because — unbeknownst to my college professors and the public at large — ludicrous portmanteaux are the origin of my entire creative oeuvre. The jig is up!

About a year ago, I was going through a big tech and web-culture phase, reading Boing Boing all the time and learning ActionScript. I began incorporating microcontrollers into the things I made. I listened to Kraftwerk. With all this tech and web and synthesized music around me, it was only a matter of time before "Moog + blog = Bloog" popped into my head.

The Bloog is the most complicated portmanteau project I've done so far. It's a synthesizer that uses not oscillators, but RSS feeds, one that manipulates not sounds, but words. The look and functionality come from 1960s-vintage Moog modular synths, but the material and results are 100% Web 2.0. This machine allows you to scramble the words and sentences of the feeds it pulls in, recycling the old material

and "synthesizing" new blog posts. The knobs scroll through the text of each entry and the sliders change the number of words appearing onscreen.

At its core is a MAKE Controller (makershed.com), which sends analog signals from the potentiometers out to a custom Flash movie on a nearby PC. The PC pulls in the RSS feeds and outputs the text to the Bloog's display (an old HP handheld).

Though this configuration was nicely functional, I realized there was something missing — beautiful lights! Since my design was based on Moog synths in only the laziest possible way, I figured it wouldn't do too much harm to add a line of LED "white-space warning lights" next to the display. They illuminate first in green, then reach a red zone if the operator has reached a dangerously low level of verbosity, a killer in today's blogosphere.

The Bloog is the first in what I hope will be a long line of preposterous wordplay-based machines. I have them catalogued on my website under "useless" right now, but perhaps I should come up with a better term. Portmantool? Mechanteau? Portmantraption? OK, I'll stop.

At this very moment, Andrew Haarsager is either making up more useless objects or writing about them at nosmarties.com.

Photograph by Andrew Haarsager